Development Communication for Agriculture

R.K. Samanta
Editor

B.R. Publishing Corporation
Delhi - 110 052

Published by:

B.R. Publishing Corporation
425, Nimri Colony, Ashok Vihar, Phase-IV
Delhi-110 052
E-Mail : *brpc73@gmail.com*

2021

ISBN 9788170186007

Printed at Balaji Offset, Delhi

Cataloging in Publication Data--DK
Courtesy: D.K. Agencies (P) Ltd. <docinfo@dkagencies.com>

Development communication for agriculture / R.K. Samanta, editor.
pages cm
Contributed articles.
Reprint. Originally published: 1990.
Includes bibliographical references.
ISBN 9788170186007

1. Communication in agriculture--Developing countries. 2. Agriculture--Technology transfer--Developing countries. 3. Agricultural extension work--Developing countries. 4. Agriculture--Research--Developing countries. I. Samanta, R. K. (Ranajit Kumar), 1948- editor.

LCC S494.5.C6D48 2021 | DDC 630.91724 23

Dedicated

to

my loving sons

Soumyajit and Subhojit

Foreword

Agricultural development is becoming an increasingly complex phenomena, particularly in developing countries which have to produce more and more food and other commodities from less and less land due to the growth in population. Agricultural scientists, extension workers and developmental administrators will have to combine efficiency and economic viability with equity and ecological sustainability in technology development and dissemination. Most developing countries are beset with problems caused by dimunition in the biological potential of the soil as well as growing biological imperishment due to the loss of biological diversity. Both biotic and abiotic stresses are increasing. It is in this context the development communication assume critical importance. Only a systems has approach in agricultural development will enable us to face successfully the emerging challenges on the farm front.

Dr. R.K. Samanta has put together a number of very valuable papers dealing with different aspects of development communication. This book is a timely contribution. We are now fortunate to be in the midst of an information age where new technologies of communication are becoming rapidly valuable. I am confident that all interested in sustainable agricultural development will welcome this important book. Our gratitude goes to Dr. Samanta for this Labour of Love.

M.S. Swaminathan
Honorary Director
Centre for Research on Sustainable
Agricultural and Rural Development

Madras, India
October, 1989

Preface

Technological innovation has been a key element in the growth of Agriculture throughout the world. But professionals in agricultural development are gradually realising that modern agricultural science and technology have a certain bias which causes a different impact on development in different regions and areas. Due to the growth of the population the low price for agricultural produce, structural changes, nomadic and sedimentary agriculture as well as apparent short-term success of introduction of modern technologies and its appeal to study the more professional systems no longer seem to satisfy the needs of the people. Therefore, there is an urgent need to develop a holistic/integrated approach to combat the problems of agricultural production and productivity and find out viable solutions to satisfy the various needs of the people of the developing countries around the world.

The present volume is a modest attempt to address these aspects with wide exposures as experienced, observed and perceived by various scholars and experts from various countries of the world. The volume is divided into four parts viz. i) Development Communication—a Systems Approach; ii) Information Systems and Agricultural Development; iii) Research-Extension-Farmers Linkages—a New Thrust for Development; and iv) Training and Technology Transfer; consisting of eighteen papers contributed by several scholars in the field.

The volume is intended to serve the dual purpose i.e., academic as well as applied. The book is expected to be of much help to the Teachers, Students, Researchers, Development Administrators, Planners, Communicators and Technology Transfer Personnel concerned to agricultural and rural development of the developing and developed countries.

I would like to express my grateful thanks to Dr. M.S. Swaminathan, former Director General, International Rice Research Institute, Philippines and presently Honorary Director, Centre for Research on Sustainable Agricultural and Rural Development, Madras, India, who has kindly contributed a Foreword to this volume.

My wife, Sarada and my sister Bapi supported me greatly in this endeavour and they deserve my full gratitude.

At the end, I would also like to express my special thanks and profound sense of gratitude to all the learned contributors who by contributing their valuable papers, enabled me to place this volume before the community concerned for the development of agriculture and human beings.

Hyderabad, India
April, 1990

R.K. Samanta

Contents

Contributors

1. Awa, Njoku, E. : Asso. Professor, Deptt. of Communication, Academic 1, Cornell University, Ithaca, NY, 14853-5901, USA.
2. Bawden, Richard : Dean, Faculty of Agriculture, University of Western Sydney, Hawkesbury, Richmond, New Southwales, Australia.
3. Bhattacharya, Subir : Research Associate, Institute of Agriculture, Viswa Bharati University, Shantiniketan, West Bengal, India.
4. Bowen, Blannie E. : Rumberger Professor of Agriculture, The Pennsylvania State University, University Park, PA, USA.
5. Carey, Harry, A. : Professor of Agricultural Extension, The Pennsylvania State University, University Park, PA, USA.
6. Chatterjee, P. : Programme Officer, Lab to Land Programme, Zone - II, ICAR, B.C.K.V., West Bengal, India.
7. Chattopadhyay, S.B. : Ex-Vice Chancellor, B.C. Krishi Viswavidyalaya, West Bengal, India.
8. Dasgupta, D. : Head, Agricultural Extension, Institute of Agriculture, Viswa Bharati University, Shantiniketan, West Bengal, India.
9. Forest, Laverne, B. : Professor, Continuing and Vocational Education, University of Wisconsin, Madison, Wisconsin, USA.
10. Ghosal, S.L. : Director, Field Extension, Directorate of Extension, Ministry of Agriculture, Govt. of India, New Delhi, India.
11. Halim, Abdul : Professor of Agricultural Extension and Director, Graduate Training Institute, Bangladesh Agricultural University, Mymensingh, Bangladesh.

12. Huque, Md. Mahfuzul : Asso. Professor, Agricultural Extension, Bangladesh Agricultural University, Mymensingh, Bangladesh.
13. Islam, Md. Monirul : Asso. Professor, Agricultural Extension, Bangladesh Agricultural University, Mymensingh, Bangladesh.
14. Islam, Md. Shahidul : Asst. Professor, Graduate Training Institute, Bangladesh Agricultural University, Mymensingh, Bangladesh.
15. Minnick, D.R. : Training Specialist and Head, Technology Transfer, International Rice Research Institute, Manila, Philippines.
16. Nkhungulu, I.C. : Asst. Director (Extn.), Ministry of Agriculture and Cooperatives, Republic of Zambia, Lusaka, Zambia.
17. Ostman, Ronald, E. : Assoc. Professor of Communication, Cornell University, Ithaca, NY, USA.
18. Patil, Jaya : Research Associate, Information Faculty, National Institute of Rural Development, Hyderabad, AP, India.
19. Ray, G.L. : Professor of Agricultural Extension, B.C. Krishi Viswavidyalaya, West Bengal, India.
20. Russel, David : Director, Graduate Programmes, Faculty of Agriculture, University of Western Sydney, Hawkesbury, New Southwales, Australia.
21. Samanta, R.K. : Associate Professor and Head, Transfer of Technology Systems and Policies Unit, National Academy of Agricultural Research Management, Hyderabad, India.
22. Singh, Ranjit : Joint Director, Communication, Punjab Agricultural University, Ludhiana, Punjab, India.
23. Tipa, E.V. : International Rice Research Insitute, Manila, Philippines.

Introduction

Technology Development for Sustainable Agriculture

The application of modern agricultural technologies which depend heavily on the use of external inputs requires a certain minimum rural infrastructure to allow supplies and maintenance of external inputs and services assumes capital investment at farm level and needs a stable environment.

Professionals in agricultural development are gradually realising that modern agricultural science and technology have a certain bias which causes a different impact on development in different regions and areas. Due to the growth of the population the low price for agricultural produce, structural changes, nomadic and sedimentary agriculture as well as apparent short-term success of introduction of modern technologies and its appeal to study the more professional systems no longer seem to satisfy the needs of the people.

Agricultural Scientists generally speak of "low resource areas" for those areas which have low potential for the efficient application of high external input technologies. As a consequence, the production potential of these areas are regarded as low and as a result, investments made for the development of agriculture in these areas have been laid behind for those made for the so-called high potential areas.

Production potentials of most tropical areas have been assessed on the basis of "Western Agricultural insights". Investments in agricultural development have been based on these assumptions and as a result locally available resources with which Western scientists are unfamiliar have been utilised.

Agricultural resources and extension services, institutions for agricultural education and inputs supply agencies, in general are based on the model and high external input agriculture and have little experience with low external input and sustainable agriculture.

There is a need now therefore, for a new practice in technology development and subsequent need for restructuring of the services, institutions and agencies involved in it. Existing entities will not change their methods and/or their systems. To bring these desired changes, convincing arguments with successful examples and operational approaches are needed. Only then, there is a chance that the entities will develop qualities such as

— a positive attitude towards existing diversity in agricultural systems
— eye in productive potentials of indigenous resources and local knowledge systems
— importance attached to participation of the population in technology development.

Subsistence Agriculture

The steady increase of the agricultural population is resulting in greater numbers of people being involved in subsistence farming. This is occurring even though many agricultural workers are entering the wage economy.

Historically, women and men have worked side by side in agriculture. The sharing of this responsibility is now being upset by new patterns stemming from male migration, commercial agriculture, and cash cropping. Since subsistence farming continues to be important for much of the rural population, non-formal education is seen as particularly useful in focusing, for example, on the production of basic food crops and on household management.

As men shift to cash-crop production, the tendency is for women to take on more of the subsistence production. Women have to fulfil this responsibility in addition to their traditional roles in agriculture and the family. Women's responses have therefore tended to evolve in two ways—as specific preferences for particular types of crops and for certain lands. Both responses have serious ecological and nutritional implications.

Although many planners recognize the importance of women's contributions to agriculture, they are often inclined to see women only as the beneficiaries of the results of programmes aimed at men. Since men predominate in development planning in general, and in agricultural planning in particular, they may tend to focus on male farmers. Often such programmes introduce male farmers to new technologies that bring them into the cash economy, but ignore or romanticize the work of women in feeding and caring for their families.

Shifts from subsistence to commercial agriculture are seen as a process by which productivity may be increased and opportunities for employment and income heightened for agrarian populations. But the process can also produce serious social, economic and environmental problems. Commercial agriculture often means specialization in a single crop or in a few crops on a regional or national basis. This can lead to narrow and conflicting interests that neglect broad national and international agricultural development aims.

Efforts to intensify production by individual farmers have met with mixed results. The production of hybrid seeds and improved fertilizers have combined with the availability of farm machinery to greatly increase the crop yields in certain regions. But these successes sometimes brought with them other problems. Small farmers who could not afford the new technologies were at a disadvantage. A concentration of land ownership has accompanied the trend towards larger mechanized farms and the subsequent displacement of small holders and many agricultural workers.

Appropriate and intermediate technologies now serve as the focus of discussion of techniques aimed at improving agricultural production. By "appropriate technology" is meant those techniques and approaches suited to particular social and environmental conditions. "Intermediate technology" implies the use of tools, machinery and ideas to supplement rather than displace human effort. In both interpretations, non-formal education for appropriate technological innovation seeks to foster local income, employment, the growth of domestic technology, and, most important, living in harmony with the physical environment.

Agricultural Extension and Communication

Many considerations must be taken into account when developing extension or agricultural communication programme. Many programmes have been developed with the primary consideration being the technical content—conducting agricultural research. A common assumption is that if we can develop an important variety of rice or new cultural practice, then farmers will automatically accept them. This so called "trickle down effect" normally does not work. First of all they have generally, fairly complex stages of adaptation needed before research can actually be put into practice by farmers. It also shows the two-way flow of information between these levels which is essential to ensure that the technology is relevant to the farmers who are the users.

The client of extension and agricultural communication programme is the farmer. However, in many cases agricultural programmes are designed with little knowledge of the farmer and his family. A farmer is often blamed for not accepting new practices recommended by agricultural extension services and are accused of being ignorant, conservative and traditional. But in actual cases, this is not the fact. Problems lie with the programme and the technology.

The primary consideration of the family is survival and so they can't take risk that might mean no food. The farmer family also experiences peer group pressure from their neighbours and friends. Additionally, there are other cultural and religious factors that influence the decision making process and that can effect the cost/benefit calculations of the farmer.

One other major reason that recommendations are often not adopted is caused by the unavailability of necessary supplies such as seeds and credit. Extension services and communication programmes are often guilty of recommending a new practice before the necessary supplies and inputs are actually available. Obviously, this weakens the credibility of extension and communication programme.

A rural based development project involves a very complex "social system" with many groups that can influence its planning and implementation. Commonly, extension and agricultural communication workers think of their job as simply relating to the message to a farmer. Further, they must be aware of the total system that can effect their programme. It can be observed that in many developing countries, there are serious communication problems between their agricultural research and their extension programme.

Agricultural extension workers have to work with many groups. They must function within the extension services organisation and be accountable to their supervisors and their superiors. They must be trained in proper manner to cater to the needs of the farmers.

Extension and agricultural communication programmes must be integral part of the central planning and implementation of rural development based development projects. If a rural based development project is to be effected, it must be periodically reviewed and modified to feed the other changing needs of the clientale and new developments in areas such as the technical and political.

Extension and agricultural communication specialists have a complex job. First, they must find compromise between the needs of farmers and the requirements of the policy decision makers. Second, the complex organisational system that they work in must be understood. Third, they must work closely as a team with people who are responsible for the other inputs required by the overall programme. Fourth, they must have an understanding of the overall programme planning and implementation strategy as well as being professionally competent at their work.

Training and Technology Transfer

For motivating farmers specially, small farmers to adopt new technology, practical training is a must. This training is also required for the extension workers those who will facilitate technology transfer among the farmers. Training of farmers through organised course can be considered to an intensive training activity which is carefully planned and properly implemented with a view to educate the participants systematically and thoroughly about a pre-determined subject matter. A successful training programme is that which can lead the training from one stage to the other. The guiding principles for working out training strategies lie in these dynamics of development process.

Organising the training will depend upon certain peculiar pre-requisites that determine the possibilities whether or not training courses can be held effectively. Training is the acquisition of the art of utilization of knowledge and skills. Effective training is not a process of cramming information into the heads of the trainees but consists of a real interaction and understanding between trainees and trainers. It is to be kept in mind that true learning takes place only if a learner feels some change after his/her exposure to a learning experience.

Development Communication:
A Concept for Development through a Holistic Approach

While there has been critical discussions and lot of writing about the problems surrounding current and past approaches to rural development, till to date, no new conceptualisation of effective communication process and strategies have been offered which appraised those problems. Rogers and Shoemaker (1971)[1] noted a need for a wider roles for communication in development, for field experiments, and for putting the communication scholar front and centre as a development designer. Another scholar sizes up the problem saying that while there has been areas of "existing experimentation" at the grassroots level, the outcome has been no more a strong conviction that communication is critical to development efforts. He suggested that there is a desperate need for communication theory formulation and research and that the history of development communication as a strategy for technology transfer in developing countries is characterised at its inception by an utter lack of theoretical conceptualisation, (Haberman, 1986).[2]

There is an expressed need for more experimental research not only to formulate theory, but to validate and motivate communication strategies which have been found to be effective and facilitating dialogue as well as insights into participatory approaches. Scholars have been talking about "participatory model," "indigenous knowledge systems," "input from the grassroots" with clearly defining concepts or inter-relationships of concepts which are a foundation for research.

The most utilised model for development communication has been Roger's (1983)[3] "diffusion of innovation model," a "top down" or "trickle down model" which has come to be severely criticised. In fact, Rogers has been critical of his own model used in research and in the latest edition of his book (Rogers, 1983)[4] discussed its limitations to propose a convergent model. Rogers acknowledges the pro-innovation bias in his early model. The later convergent model (Rogers and Kincaid, 1981)[5], recognized the importance of dialogue between proponents and participants, providing an exchange of information and experience in order to reach mutual understandings. They said that communication is a two way process of convergence (or divergence) rather than a one way linear act in which one individual seeks to transfer a message to another. Later they recognize that innovation can be diverted indigenously and acknowledges that the social structure is a strong determinant of who can benefit from innovations exist externally. It seems clear that the social system is a critical factor in access to that social learning becomes an absolute necessity if individual takes advantage of that which has accessed.

The model of development communication, derived from the division of innovation paradigm, expects the target group to trade comfortable old practice for new uncertainty on the basis of recommendations from external authorities or experts. Rather than a one way persuasion process it is recognized that adoption is more likely to be successful when the potential adoptor is involved in a dialogue about their own needs and available alternatives. Thus, there is a need for a new paradigm which would reflect their dialogue.

In the past, participation was interpreted as a villager's behavioural response to persuasive stimuli, urging the adoption of new practice. Feedback has frequently been used as "devised to make external manipulation more effective". But not as a mechanism for making intervention agencies more responsive to the needs of the community (Kearl, 1978).[6]

Cohen and Uphoff (1980)[7] have elaborated a framework to clarify the concept of participation in rural development. They suggested three dimensions of participation, — what kind of participation taking place who is participating and how the process of participation is occurring

They pointed out the kinds of participation that warrant major concern; participation in decision making; participation in implementation; participation in benefits; and participation in evaluation. With this changed emphasis on participatory rural development, the perception for development communication will of necessity need to be adjusted to accommodate the change. Subsequently, the role of the development communication will change. Definitions of development communication are many, ranging from the simple one said by Venkataraman (1983)[8] that development communication is primarily communication for planned change. The overall thrust of development communication is to substantiate the idea of agricultural research and extension as a process rather than a discrete or mechanically aggregated activities. It is argued that technological know-how can best be incorporated into the agricultural production process if extension is used to build a continuum between research and its ultimate beneficiaries, the farmers.

Keeping all these in view the present volume has been planned to incorporate various aspects and dimensions of development communication for all round development in agriculture as experienced and observed by various scholars and experts from the concerned fields. The Volume has been divided into four Parts namely: (1) Development Communication : A Systems Approach; (2) Information Systems and Agricultural Development; (3) Research-Extension-Farmers Linkages—A New Thrust for Development; and (4) Training and Technology Transfer.

Each chapter of each Part expresses a detailed picture of the problems and its possible solutions in an integrated manner incorporating various facets of the concerned theme. Each Part has highlighted the problems, importance of its study and the probable solutions to overcome these problems in a holistic pattern considering the varying needs of different clientale of different walks of life, particularly, farmers, extension workers, agricultural research scientists, development communicators, agricultural administrators, policy makers and overall, social scientists. It is expected that this Volume will meet the long felt needs of the various groups of these users and hoped that the book will help promote the cause of agriculture and rural development and will pave the way for further research, investigation and discussion into various aspects of rural habitat, technology, economy and society.

R. K. Samanta

References

1. Rogers, E.M. and Shoemaker, F.F. (1971), *Communication of Innovations: A Cross-Cultural Approach*. The Free Press, New York.
2. Haberman Peter, (1986), *Back at the Drawing Boards: The Third Decade in Development Communication*, paper at the 36th Annual Conference of the International Communication Association in Chicago.
3. Rogers, E.M. 1983. *Diffusion of Innovation* (3rd Ed.), The Free Press, New York.
4. Ibid.
5. Rogers, E.M. and Kincaid, D.L. (1981). *Communication Networks: Toward a New Paradigm for Research*. The Free Press: New York.
6. Kearl, Bryant E., (1976). "Communication for Agricultural Development," in Shramm, Willbur and Learner, Daniel (eds.). *Communication and Change : The Last Ten Years*, University Press of Hawaii : Honolulu, Hawaii.
7. Cohen, J.M. and Uphoff, N.T. (1980). "Participation's Place in Rural Development: Seeking Clarity through Specificity," *World Development*, Vol. 8, pp. 213-233, Pergamon Press: London.
8. Venkataraman, A. (1983). "Development Communication and Grassroots Participation in Food Production, India", in Ouchi and Campbell (Eds.), pp. 159-171, Association of Development Research and Training Institutions of Asia and the Pacific (ADIPA).

Part 1

Development Communication —A Systems Approach

"Development is a widely participatory process of social change in society, intended to bring about both social and material advancement (including greater equality, freedom, and other valued qualities) for the majority of the people through gaining greater control over the environment."

E.M. Rogers, 1976.

1

Systemic Action Research—Extension Re-Constructed

Richard Bawden and David Russell

Introduction

Technological innovation has been a key element in the growth of agriculture throughout the world (Dorner, 1983). Indeed as Mellor (1986) claims "....one of the most important theoretical and empirical findings in analysis of (all) western economic growth is the identification of technological change as a major source of growth".

Growth in levels of agricultural production and productivity have not been confined to western economies. Many developing countries too have taken advantage of international initiatives to restructure their research, extension and educational services, with very positive responses being seen in their agricultural sectors. Some of these advances have been so spectacular that terms like 'green revolution' and 'quantum leaps' have been used to describe them (Jain, 1988).

Yet in spite of such apparent progress in food production, most developing countries still face overwhelming problems "...of growing population, recurrent food shortages, illiteracy, malnutrition, environmental pollution, and economic and social disparities" (Anon, 1986). Chambers (1983) puts it more bluntly "...the extremes of rural poverty in the Third World are an outrage".

Many have argued that it is simply a matter of transferring technologies and the institutions which develop and promulgate them, from "developed" nations to those still "developing". Others suggest that such a technology-focused approach is singularly inappropriate and that a new "people-focused" framework for development, is critical (Korten and Klauss, 1984).

While this switch in emphasis is appealing at first sight, it is still inadequate for it is really impossible to separate technology from people: Each is inexorably connected with the other. Therefore it is not a matter of technological innovation for productivity versus increased human welfare through human resource development. Both are important and neither is adequate without the other. There is here what Bertalanffy (1968) refers to as a glorious unity of opposites. Rather than being different issues, they are different faces of the same issue, where the issue itself is different again from its two conventional faces: The whole is different from the sum of its parts (Bateson, 1979). Improvements in any current situation will come about through the exploration of the nature of the differences between the two positions, viz. technology-centred and people-centred, rather than the substitution of one approach for the other. Furthermore, such an exploration will need to address rather than ignore, or merely reduce, the ambiguities and complexities of rural development (Bawden, 1989a).

Such a shift in focus will demand new theories as well as novel practices. It will also bring into question profound philosophical issues of ethics, values and morality, as well as notions about ways of knowing, meaning and knowledge. Of particular significance here is the emerging paradigm of systems thinking and practices (Bawden, 1989a).

In this chapter we shall further explore some of the grounds for considering changes to the existing frameworks or paradigms of agricultural and rural development. Secondly, we will examine some of the foundations and limitations of the conventional models of the classical institutional trinity of research, extension and education. Next we will focus particularly on communication as a vital process in providing critical linkages for development. Finally, we will develop an argument for systemic action research (SAR) as an alternative paradigm for development. Many of the ideas we present in this final sector will be based on experiences in our own College where, for more than a decade and in concert with others, we have been actively researching and developing such a paradigm (Macadam and Bawden, 1985).

Agricultural Development : Rural Undevelopment!

Given the dichotomy of finite resources on the one hand and ever growing needs for food on the other, the virtual obsession of agricultural developers with the concept of productivity, is understandable. Improvements in this parameter of production efficiency are to be found as the central claim of agricultural development, throughout the literature (Kingma, 1985). Further yet is the tendency to ascribe the vast majority of productivity growth in agriculture to technological innovation and to the scientific research which spawns it (Bonnen, 1983).

At best, this centrality of productivity as the parameter for agricultural development, is naive and simplistic: At worst it can disguise, at least temporarily, serious long term degradations in the bio-physical and/or socio-cultural rural environments in which agriculture is conducted. Left unaddressed, agricultural development might result eventually in irreversible rural undevelopment!

Another way of putting this is that strategies for improving agricultural productivity are often (and increasingly?) associated with undesirable environmental outcomes—negative externalities as they are often termed. In the USA "...these issues include (bio-physical) environmental degradation; concerns for animal welfare; impacts on health and safety of farmers, agricultural workers and consumers; adverse nutritional effects of production and processing technologies; the extrusion of smaller family farmers from agriculture, erosion of rural communities, and the concentration of agricultural production and economic wealth; inadequate conservation and commercial exploitation of fragile lands that should not be in cultivation" (Bonnen, 1983).

Soil erosion, desertification, chemical pollution, salinization and acidification are all examples of degrading and widespread environmental "side effects" of agricultural production technologies (Bawden and Busch, 1988). Reduced genetic diversity, biocide resistant strains and varieties of pathogens and pests and parasites, and changed weed profiles favouring the invasion of woody shrubs, are other examples of biological environmental stresses associated with perturbations of agricultural practices. All have appeared as undesirable if unintended outcomes of the singular quest for productivity growth. And where these once might have been but local issues, there are some extremely disturbing signs of their contribution to macro-global degradations such as the so-called "greenhouse effect". There is no doubt of the global impact of some of the economic aspects of growth in agricultural productivity. Wolf (1982) cites examples such as wheat, rice, sugar, bananas, coffee and even opium as examples in support of his thesis of profound historical socio-economic changes in the face of changed production patterns and levels. These include induced unemployment and even trans-national emigration.

Aggregate figures for improvements in agricultural productivity for whole sectors or commodities or industries frequently disguise, indeed often comprise, individual farm failures.

There is something fundamentally self-limiting about productivity growth whether it be measured solely in terms of physical yields or economic efficiency. One of the paradoxes of the industrialisation of "advanced" agriculture is that as farming systems change from pre-industrial to fully industrial, as measured by the intensity of capital inputs, their productivity in terms of energy efficiency, declines very significantly (Bayliss-Smith, 1982). In viewing productivity as a central focus for development there is a need to combine a parameter of the farming system itself with measures of the environment in which it exists: As Kingma (1985) has noted "...the attempt to evaluate productivity change using economic efficiency only, is to address only part of the issue (for) productivity increase influences the social conditions of production, and vice-versa".

Conway (1987) addresses this situation by proposing productivity as a measure of social value "...the output of valued product per unit of resource input". In so doing, he defines social value as "...a function of the amounts of goods and services produced by the agroecosystem, their relationships to human needs, or happiness and their allocation among the human population". Conway himself, has proposed extending measures of agricultural development beyond productivity to also embrace stability, sustainability and equitability. Rambo (1985) further extends these measures to include solidarity and autonomy.

The crux of this perspective is that if agricultural development is to be congruent with sustainable rural development then it must also focus on the increasing welfare of rural people on the one hand, and with the maintenance of the integrity of the bio-physical environment on the other.

These multi-dimensions are a crucial part of this debate for they set a new context for processes such as extension and research. This in turn will dictate the need for new strategies, policies and practices.

New Institutional Directions

Technological innovation has been at the heart of agricultural development from the time of the first digging stick! Through trial and error, new husbandry practices were invented and shared in such a way that they "diffused" across vast areas of the globe, from early times. Technology, which in some cases like irrigation systems were highly sophisticated, long preceded scientific research in its application to better agricultural productivity. It really was not until the middle of the nineteenth century that scientific research and technological development for agriculture were integrated and institutionalised. In the United States this was combined with education at the same time and eventually also with extension as a formal process of transferring technology. This is worth pursuing, for much of what is done with these three functions worldwide, is a variation on the model which had its foundations in the middle of the nineteenth century. It is our belief, that much of the intellectual tradition associated with these foundations, continue to this day.

Although it took some forty or fifty years to be consolidated, the institutional "trinity" of functions of teaching, research and extension was "born" with the birth of the so-called Land Grant Universities in the USA. The passing of the Morrill Act granting their establishment in 1862, was coincident with extremely significant scientific and technological developments a continent away, in Europe (Bawden, 1988). Thus the eighteen sixties were neatly sandwiched between those two monumental works of Charles Darwin—"The Origin of Species" and "The Descent of Man". This was also the era, albeit unrecognised at the time, of Gregor Mendel's genetic experiments. In addition to the powerful insights into the "biology of the world" as provided by Darwin and Mendel and others, there was a burgeoning of scientific discovery in other disciplines relevant to agriculture by such "giants" as Louis Pasteur in France, John Lawes and Henry Gilbert at Rothamstead in England and Justus von Liebig in Germany. The latter established a tenet of fundamental importance to subsequent generations of agricultural research scientists when he wrote "....the growth of a plant is dependent on the amount of foodstuff which is presented to it in minimum quantity"; find the minimum quantities of each nutrient, and supply it and move on to the next limiting factor!

This scientific explosion provided not only new knowledge for the foundations of new technologies, but new ways of knowing for the foundations of research methodologies themselves. Whether one talks of Baconian experiments of the fact gathering kind (Mendel) or of Galilean experiments which test hypothesis or allow discriminations between possibilities (Darwin), to refer to distinctions made by Medewar (1979) such scientific methods rely on particular "ways of knowing".

Implicit in the scientific tradition established at this time and adapted in the main by agricultural institutions like the Land Grant Universities in the USA are the notions of: reductionism (the whole can be explained through investigation of its parts); mechanisms (all natural phenomena, including life itself, allow mechanical explanation by physics and chemistry); positivism (recognising only positive facts and objectively observable phenomena); and rationalism (there is a reasoned and correct explanation for everything).

The initiation of the Liebigian paradigm of incremental improvements in biological productivity through the isolation of the "next limiting factor" was paralleled by the development in social science of the concept of the economic margin. Neo-classical economists like Stanley Jerons, Carl Menger and Leon Walras began to integrate notions of utility and marginalism into the classical economic models of Adam Smith, Malthus and Ricardo. Agricultural research was thus born at a time when key concepts of incrementalism in both the natural and the social sciences were being articulated. It was also the time when the foundations were being laid for the empirical, and "apparently" value-free or objective ways of inquiry which have become refined as the scientific method.

Equally of course, the latter half of the nineteenth century was a period of enormous social change in virtually every corner of the globe through civil war, revolution and colonization. As the industrial age continued its development it spawned further alternative socio-economic models to the neo-classical and classical ones. Karl Marx, Friedrick Engels and Herbert Spencer were all providing ideas which have profoundly influenced the course of many events, not least agriculture and rural development. A huge proportion of the world's agriculture is currently conducted within centrally planned economies whose political economies are based on the ideas of Marx and Engels. And the influence of these men is certainly not confined merely to political economy. Indeed it has been said that "...underlying all of Marx's mature writings is the search for social reality" (Coser, 1971). Although mainly eschewed by those who have been the key actors in the development of agriculture in western nations over the past one hundred years or so, some of the crucial concepts of Marxism are beginning to appear as the focus shifts towards development models which stress the centrality of human resources. Of particular significance in this regard are the critical theories of Jurgen Habermas (1971) and their place in the development of participative action research paradigms.

Objectivity in Science : Its Influence on Extension

Many now recognize that agriculture is not merely a technology, but a human activity which sees technology being used in physical and socio-cultural environments (Bawden and Busch, 1988). This perspective provides new challenges for agricultural science, extension and education, not the least because it must question the conventional view of the objectives of the world and traditional problem-solving strategies based on linear models of knowledge generation, transmission and adaption! Herbert Spencer was one of the early social scientists who addressed the special problems of objectivity that arise in the scientific investigation of a social world in which the researchers themselves take part. He was particularly interested in the similarity between social and biological organisms and stressed the importance of language which allows a permanence of relations between individuals within particular societies. Of particular interest is his belief that "the properties of the units determine the properties of the aggregate" and that "every phenomenon exhibited by an aggregation of men originates in some quality of man himself" (Spencer, 1891). This tradition continued to pervade sociological thought for many decades and has had profound effects on institutional development—not least that of extension. As Fliegel and von Es (1983) record it, this "intellectual tradition which defined sociological phenomena at the micro level and explained behaviour in terms of individual acts" dictated the direction of research and extension as the introduction of new technologies at the farm level right through until about the nineteen sixties.

It was this view that supported the model of research→ knowledge→ transfer→ adoption→ diffusion. Reflecting the spirit of incrementalism, so much sociological research has gone into "individual barriers to adoption" while the model itself remained virtually unchallenged. This has been all the more surprising given that, in practice the model has been found wanting now for almost twenty years (Roling, 1985). "Simply put, the model, which had its theoretical and conceptual origins in the developed western world, has proven irrelevant to the problems that confront the vast majority of rural people in the Third World who depend on agriculture for a livelihood" (Oakley, 1988). In spite of this there are few signs of conceptually attractive improvements to the prevailing model; "The criticism and disenchantment towards the diffusion of innovation approach in research and extension application have highlighted the growing awareness of the shortcomings of diffusion theory which have paralleled the accumulating evidence on the low effectiveness of extension services in various countries" (Crouch and Chamala, 1981).

A decade or so back, an alternative approach to development beyond technological innovation (TI) began to emerge. In addition to the obvious empirical evidence that TI just was not working in many circumstances, came: (a) new concepts and theories of the process and context of rural development; (b) new philosophies and methodologies of rural sociology; and (c) new concerns about the serious degradation of the environments in which many agricultures were being practised. In essence, the shift in focus has been from productivity growth through the adoption and diffusion of technological innovation, to the sustainable and equitable welfare of rural people (Coombs and Ahmed, 1974) (or human resource development) particularly through education, organisation and conscientisation (Roling and De Zeeum, 1983). On-farm micro-issues and individual behavioural responses of the farmer-as-actor, have given way to macro issues such as the effects of new technologies on the environment beyond the farm and how these are being dealt with, and the effects of capital intensive innovations on the very structure of agriculture itself. This new focus has prompted Buttel (1985) to distinguish between production sciences with their mission of production enhancement and impact sciences. They differ in both philosophical and methodological postures, and there is a certain urgency about their balanced development given the emergence of powerful new technologies based on genetic engineering.

As a number of writers have concluded, there are both enormous opportunities for biotechnology, as the next "revolution" for agricultural development, and very significant socio-economic threats (Hansen, *et al.*, 1986): "...Biotechnology has already set in motion a major restructuring of the agricultural sector" (Kloppenburg and Kenney, 1984), and not the least of these changes is the increasing "privatization" of both research and extension. Private ownership of technologies has been in existence since copyright laws were first enacted. What is much more recent is: (a) the ownership of knowledge in the form of patentable intellectual capital; and (b) the patent rights to plant and animal varieties. These factors present serious constraints to the free flow of information across communication networks. It also raises extremely pertinent questions with respect to science and human values; an issue of particular relevance to new thoughts on "development" (Swami Ranganathananda 1985).

Communication and Extension

Extension has been envisaged as a process of communication since it was first institutionalised; the earliest model, research→ extension→ adoption, being conceptualised as a transmission of generated knowledge from scientist to farmer through an active agent, and its subsequent diffusion through farm communities. For many of the reasons outlined in the previous sections of this chapter, this linear model of communication is inadequate in many ways, even when confined to the context of technological innovation.

According to Oakley (1988), there are four major factors which account for the failure of the model of technology transfer to significantly influence third world development:

— there are marked differences in the heterogeneity of rural society in the third world compared to the developed world;
— the assumption that innovations are relevant to all members of a social system and that non-adaption could be explained by psychological variables or indicators such as levels of education;
— the emphasis on the identification of "key" and "progressive" farmers who would be instrumental in diffusing innovations to other farmers; and
— the belief that the causes of poor agricultural performance were essentially technological and could be solved by developing technology and improving delivery.

Early definitions and discussions, imply if not specifically state a dichotomy between research and extension and the one way flow of information (Rogers and Shoemaker, 1971). The Transfer of Technology (TOT) model casts the researcher in the prestigious role of being the source of all new technologies, and the farmer as the passive recipient, whereas the evidence is clear that this has never been the case (Davies, 1988). Communication in agriculture is better conceived as occurring through a network rather than a conduit pipe! The realisation is, that both the message (of sustainable rural development) and the media (multi-directional networks rather than uni-direction flows) are much more complex than at first envisaged.

As Chamala (1987) has posited "....understanding the communication process, delineating deficiencies and suggesting methods to increase effective communication (in agriculture) is a difficult task because many institutions, groups, organisations and end users are involved in the process of technology development and dissemination".

A broader context of the communication network in agriculture has been provided by Salmon (1980) who presents a resource model of extension which puts farmers and researchers as equally important contributors to "...a pool of agricultural knowledge which resides in and is stored by, all those related to the industry...The function of extension is to transfer and nurture this pool of knowledge within the rural industry. Thus extension embraces all those who contribute knowledge or transfer it to farmers". As farmers do this to benefit other farmers, they are thus legitimate extension workers; as much as those professionals officially employed as extension agents.

The importance of cultural beliefs and what has been termed "indigenous knowledge" has often been undervalued in the quest for more productive agriculture. A typical view is that of indigenous knowledge impeding the development of improved systems of farming—of the knowledge of indigenous people as being superficial and primitive, and grounded in myth and taboo. Yet increasingly, examples appear which give life to such viewpoints. In what is regarded as a fundamental shift in the conventional view, Conklin (1957) established that a group of shifting cultivators in the Philippines could identify 1,600 different types of plants in a locale where an earlier systematic botanical survey had revealed a mere 1,200 species. Another more recent study (Brokensha, *et al.*, 1980) revealed the details of a folk taxonomic system for potatoes which "....had an internal consistency and could be related to the genetic classification of the potato...it also provided a basis for understanding the selection of varieties by peasant farmers pointing out their primary criteria for evaluating new varieties".

A general context about the quality of the knowledge possessed by indigenous people is provided by Lovelace (1986) with hints about new ways of communicating about agricultural development "Rather than discouraging traditional beliefs, values and knowledge, it may be, in many cases, better to encourage their maintenance and continued accumulation of information within the indigenous ideatioral framework". Rhoades (1987) and Rhoades and Babbington (1988) provide considerable support for this position. Alverson (1984) too refers to the "wisdom of tradition" where knowledge gained over generations of experiences embedded in the culture which represents a source of presumptions for future generations (Bartlett and Brown, 1985). Yet Chambers (1983) also points out some of the major limitations of local knowledge particularly where it reflects restrictive ways of knowing. There is even the rather horrifying thought that, through endless familiarity, any individual or group of human beings can reach a state where they believe that they are reacting in novel ways in the face of new events, when in fact, their behaviours are totally conventional. Such mindlessness can also be characterised by the situation where someone is not aware of any potentially novel aspects in any information which he or she is processing (Langer, 1985). There is thus a need for a delicate balance between the benefits of conventional wisdom and the paralysis of mindlessness through familiarity (Bawden *et al.*, 1989).

Emerging from this discussion, is the notion of extension as a process by which effective communication networks are established: Effective in the sense that new knowledge and new ways of knowing can be shared through common language. This presents a huge challenge to the scientist and extension agent alike who are conventionally steeped in the esoteric language of science. As Fals Borda (1980) puts it, "....the creation of knowledge that comes from the people contributes to the realisation of a 'people's science' which serves and is understood by the common people". Hall (1981) has drawn from this notion, the importance of:

— returning information to the people in the language and cultural form in which it originated,
— establishing control of the work by the popular and base movements,
— popularising research techniques,
— integrating the information as the base of an action/reflection rhythm of work,
— learning to listen, and
— recognising science as part of the everyday lives of all people.

This last point is vital. According to Kelly (1955), the experimental approach to learning about ourselves and our environments is an innate characteristic of all human beings. In postulating his theory of construct psychology, Kelly pictures everyone as a scientist—actively investigating our worlds, building up meanings or constructs, and then testing these meanings against reality (or using them to change the reality!)—key to his position is the explicity reference to the connection between the theory (of construct psychology) and its philosophical underpinnings (of constructivism). The assumption here is "....that whatever nature may be, or howsoever the quest for truth will turn out in the end, the events we face today are subject to as great a variety of constructions as our wits will enable us to continue...even the most obvious occurrences of everyday life might appear utterly transformed if we were inventive enough to construe them differently" (Kelly, 1955).

This stance is far removed from the conventional, which as argued earlier, is grounded in positivism—a position which admits to an ultimate truth of the nature of reality (Reason and Rowan 1981).

Penman (1987) turns the concept of constructivism on to the nature of the process of communication itself and concludes that "....it is the means by which we come to know of the world. It is the means by which meaning, values and social institutions are constructed". Communication cannot be the transmission of information about the world—rather it is a social act by which meanings are brought forth through social interaction. It is a participative, co-operative process where all who communicate together, are empowered to be part of the creation of knowledge.

As extension is communication, then it is a collaborative process of finding new meanings about the world where all who participate in the process, contribute to the meanings and thus knowledge generation. Rather than extension being the diffusion of the results of research and extension, from this viewpoint, extension is research, albeit a somewhat unconventional form of research.

In essence, we picture extension as a process of helping farmers to learn how to become active as "researchers" themselves—how to increase their awareness of science and scientific ways of thinking and doing things. The vision is one of "learning or researching communities" with people joining together and sharing their problems in ways which allow them to "learn their way through" the complex issues they are facing. Farmers, and members of their families, are thus encouraged to accept responsibility for their own learning and development, and through co-operation and participation, develop what has been called "social learning"—where "individual learning and behaviour are shaped by the nature of the group activities in which the individual participates, and through which he is molded and which, in turn, are frequently molded by him" (Dunn, 1971).

In the tradition of Herbert Spencer, we see the importance of the individual, influencing and being influenced by those around him or her—indeed, we see the complex of human groups as "action researching systems" (Bawden, 1989b).

Action Researching Systems

Action research is a participative, experiential process where people work and learn together in ways which allow them to make sense out of their experiences as a basis for action to change them. Its foundations were laid by Kurt Lewin. It was his intention (Lewin, 1946) to move away from the limitations of traditional research which he perceived as lacking any component of self-review. Results tended to support or fail to support pre-established hypotheses so the research process had an all or nothing character to it.

Lewin proposed that the focus of research should always be open to change; in fact, by the very nature of complex events, it had to change. He sought to combine by means of a spiral of learning cycles, the dual goals of improving practical problem situations and the discovery of basic knowledge about the way people act in their environment.

What was lacking in the early applications of the action research model and what contributed to its near demise in the fifties and sixties was the failure of researchers to develop the second of Lewin's goals, namely, the discovery of basic knowledge about the way people act in their environment. Lewin was striving for the complementary development of theory and practice. He wanted strong theories that would be applicable to pressing day-to-day problems and it would be from the interaction of the two that major improvement would eventuate. The two goals of action research and the dialectical process linking them both (the change process), is illustrated in Figure 1.

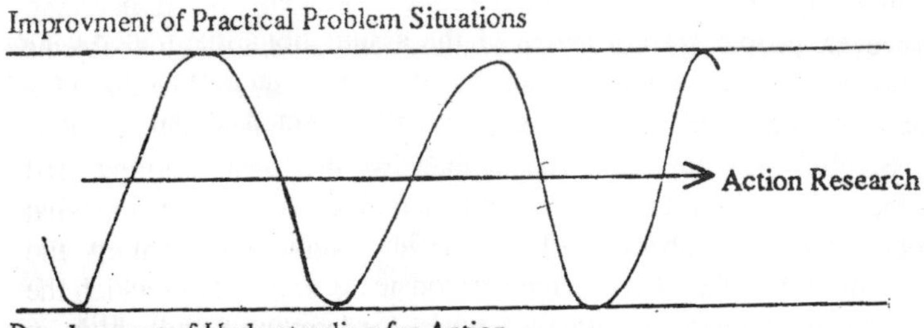

Fig. 1: The Dual Goals of Action Research.

Lewin's spiral of change is a never ending process reflecting how greater understanding of 'what is going on' influences the nature of the intervention in a positive way, which in turn, contributes to it even greater understanding. Informed reflections lead to informed interventions as seen in Figure 2.

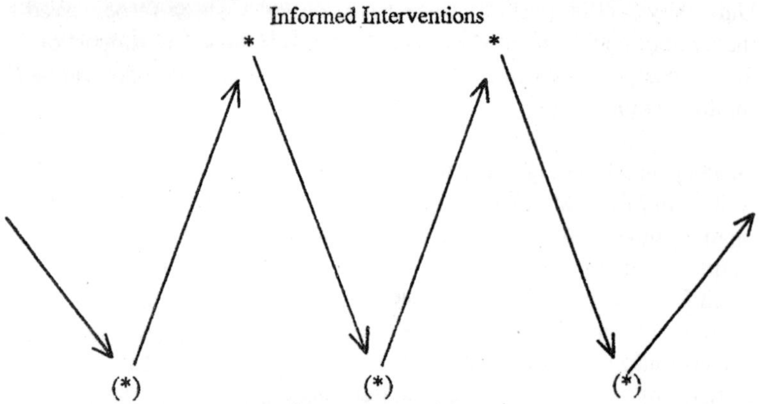

Fig. 2: The Mutual Shaping of Interventions and Reflections.

What is meant by 'informed' is that the reflections and interventions are shaped by a particular view of the world and do not take place in some sort of an epistemological[1] vacuum. Not that there ever could be such a theoryless state since every action is performed in accordance with a set of beliefs about how learning and change take place. Often, however, people are not aware of the assumptions that underlie their actions and this inevitably results in 'poor' understanding or ill-informed reflection, with consequential, ill-informed action.

Stephen Kemmis and his colleagues at Deakin University in Australia have developed their 'understanding' in the context of critical social theory (Habermas, 1972; 1974). Their particular focus is education and they have achieved considerable success over the past decade in attaining Lewin's dual goals. (See Kemmis *et al.*, 1982; and Kemmis, 1985 for an indication of the extent of this work).

There is no appropriate view of the world which will be superior to all others as a foundation for research, learning, and action. There is no absoluteness about identifying underlying premises; what we are working with in identifying where our 'understanding' is coming from, our epistemology, is a *fiction*. It is an integrated game of guesswork that seems to best fit our experience of reality.

The Hawkesbury Approach

The theoretical context for educational, research and consulting activities of the Faculty of Agriculture and Rural Development at this University for the past ten years has been the application of systems theory and models of experiential learning. It is not surprising then that action research has been seen as a likely vehicle for extending 'situation improvement' activities into the wider agricultural community.

Developing Understanding for Action

The brief history of action research has shown that to neglect the ongoing development of a theoretical paradigm which will continually shape the nature of the intervention for improvement is to sound the death knell of genuine advancement. What we would get would be a continuation of the proliferation of 'action' without understanding and our institution of learning would become a foundry for the moulding of technicians who want to solve the problems of agriculture when agriculture itself doesn't know what the problems are.

The exploratory phase of any investigation can be viewed from three levels. The first is the observation of the data (what appears to be happening); the second is a bit more abstract and consists of creating various maps, models or descriptions of the situation; and the third is still a little more abstract and has to do with arranging the data in a particular manner. Gregory Bateson described the activity of this level as a "self-conscious discussion of the procedures by which the pieces of the jigsaw are put together" (Bateson 1958). It is this final level of abstraction that expresses how our understanding is attained and it is at this level that systems thinking influences the action research model.

Systems thinking, encompassing ecological and cybernetic thinking, provides a framework for how data can be arranged and understanding developed. Discussions at this level are within the realm epistemology and it has been a constant theme at Hawkesbury that such discussions are very relevant to the achievement of desirable goals. As we are practitioners involved in bringing about what we perceive to be improvements in agriculture we cannot *not* have an epistemology. Bateson (1977) elaborates this point:

"All descriptions are based on theories of how to make descriptions. You cannot claim to have no epistemology. Those who so claim have nothing but a bad epistemology. And every description is based upon, and contains implicitly, a theory of how to describe".

The systemic epistemology that shapes the nature of an intervention is, then, a way of approaching (or knowing) problematic situations through a framework representing systems theory, ecology, and cybernetics, a large part of the educational program is designed to help participants see and know situations in a systemic way.

Systems Thinking

In our view, a system is always a short-hand way of specifying a system-environment relationship. A system can have no meaning except in relation to its context, usually termed its environment. An epistemology based on systems thinking enables one to organised the vast array of phenomena (available data) so that 'systems' can be identified. Systems thinking is best seen as a descriptive approach to the world—the ever-changing world! It wants to stay clear of generating 'laws' and anything approaching absolute models, the models it plays around with are very provisional models with the purpose of rendering observations more intelligible.

In contrast to a linear epistemology which emphasises discrete elements and properties (e.g. roles, motivations, and values), systems thinking attunes itself to inter-relation, complexity and context. The focus of action research will always be a 'system' and in systems thinking terms, a systems is "any unit containing feedback structure and competent to process information" (Bateson 1971). What becomes critical for diagnosis is knowing how the cybernetic network of matter-energy and information is interlinked or structured. Intervention then explores how restructuring might take place so as to achieve as improvement for the system.

Problematic Situations in Extension

From a systemic perspective, what is perceived as a problem is always an indication that something is not in order within the ecology of relationships within the system, or, between the system and the environment. The problem then is a message to us that there needs to be a change in the relationship network and that diagnosis and intervention should be directed towards this network rather than exclusively any isolated part.

Bradford Keeney (1979) gives an example of this when he says that:

> "The idea of seeing symptoms as communications about the ecological condition of whole systems is not new. Some members of the American Indian culture have long subscribed to an epistemology that sees systems in this way. For example, when the Navaho medicine man encounters an individual expressing a symptom, his choice of action is to treat the whole tribe in order to bring it back into harmony. In this way individuals are relieved of symptomatic discomfort through treatment of the whole ecosystem".

Cybernetic theorists such as Keeney (1983) argue that the presence of a problem indicates a system's effort to maximize or minimize a particular experience or event. The obvious degradation of a farmer's soil fits into a particular interactional context. The 'site' of the problem appears to be the soil. Yet the soil is responding to the environmental influences, including the action of the farmer, in a way designed to preserve its own stability. In order to compensate for the impact of the farmer trying to maximise a particular variable, it changes its structure and the process of gradual degradation is the inevitable result.

The challenge for the action researcher is to enter this system in a manner that fosters a more adaptive way of maintaining the organisation of the whole system. Successful intervention amounts to encouraging the system to heal itself, to 'self correct', in cybernetic terms.

Extension Officer as Researcher and Participant

What has become increasingly clear that the student(s) of systems is their own role as observers. It was the observer, in order to make a distinction, who specified that a system was a unit distinct from its background. It was the observer who then attributed to both system and background their respective properties, properties which justify the act of seeing them as separate. So, exactly what is seen to constitute a system is a decision made by a outsider who, for his own reasons, wants to explore a set of relationships.

No matter how objective the researcher wants to be, she is inescapably part of the field being observed in exploring, diagnosing and intervening in the 'problem'. The situation is the same as that demonstrated by Heisenberg with his famous 'uncertainty principle'. The principle states that the observer constantly alters what he observes by the obtrusive act of observation. Lewin challenged an accepted role of the scientist as a disinterested, 'objective' observer and this challenge is still as relevant today as it was then.

Not only does the action researcher (Extension Officer) generate the system to be studied but he is very much a part of that system that he has created. He becomes part of the network or relationships and is consequently caught up in the feedback and feed-forward (expectations about the future) processes that constitute these relationships. This is a very different picture than the one accepted by traditional science where the researcher sees herself as outside the 'problem'.

The whole system now becomes the client, the client's environment, and the researcher; it is the mutual, reciprocal and simultaneous interactions of these basic elements that constitute the whole system under investigation. Change is not then limited to the interaction of the client and her environment, but is open to the researcher as well. As the client acts on the researcher, the researcher acts on the client in a simultaneous fashion. Any lines of demarcation between the roles of client and researcher become blurred as the client becomes co-researcher and the researcher becomes an integral part of the whole system. Identification of the problematic situation can only occur through interacting with it; this was Lewin's basic premise and can be clearly seen in Figure 1 with his built-in process of continual revision of the general plan. The way the researcher interacts becomes part of the system and consequently influences whether change will occur.

Extension Officer as Researcher and Conceptualiser

We have described how the extension agent as an action researcher has become part of the network of ecological relationships that go to make up the whole system out of which the problematic situation is identified and hopefully changed. There is a real risk in this outcome that the system will adjust over time to the input of the researcher so that a state of homeostasis will prevail and change will no longer be possible. The stable state of client, problem, and environment has been replaced by a new stable state of client, problem, environment and researcher. When this happens the researcher's ability to contribute 'understanding for action' is lost.

The particular understanding that the action researcher, imbued with systems thinking, will bring to the situation will be the skill to reframe the perceived problem into a communication about the interactions within the system and/or, the system with its environment. Such a reframing will then give the direction to where and what the intervention should be. Systems thinking is the conceptual base that informs and gives meaning to the interactions within the whole system. The action researcher, as participant-conceptualiser, is the one who continually explores: (i) how meanings are being created; (ii) how they are stored, (iii) how they drive action; and (v) how they might be changed (Smith, 1983).

In order to short circuit potential blocks to the researcher's effectiveness in this area, the provision of a conceptual 'support' system is most appropriate. Such a 'horizontal expansion' has been found to be most beneficial in family therapy (e.g. Jackson and Satir, 1973; Auerswald, 1973; Selvini-Palazzoli *et al.*, 1978). The function of the interacting support system is to prevent any closing off of the system which would result in a state of unproductive homeostasis. The researcher is a part of both systems: he needs to join the client and experience himself as part of the problematic situation, and at the same time, he needs to remain part of a conceptual resource team. Remaining

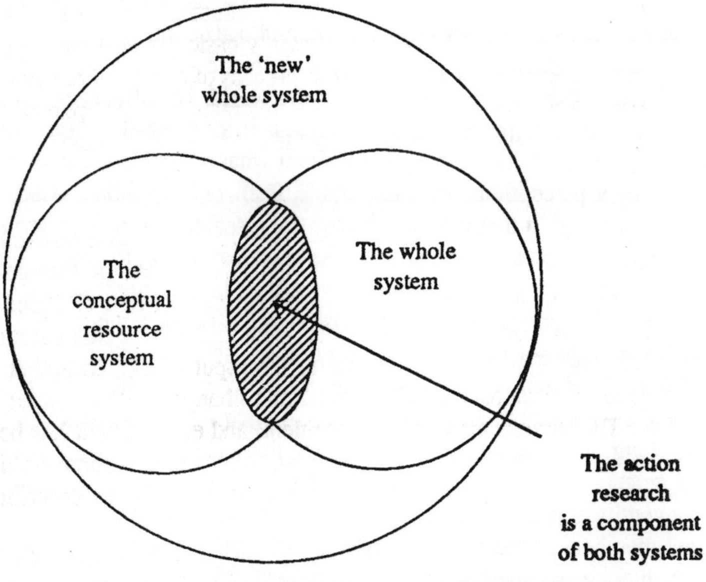

Fig. 3: The Generation of a 'new' Whole Svstem

an active part of this team enables him to gain the necessary conceptual inputs when he feels swamped because of his immersion in the client system. The inter-relationship of the two systems can be expressed as in Figure 3.

Diagnosis and subsequent intervention can now be seen as the interweaving of two systems.

Conclusion

We believe that picturing the extension agent as an action researcher, provides an important new perspective on the process of the responsible development of rural communities. Furthermore, we posit that the model of action researching systems is most appropriate for today's world. It has evolved in a period in history when the achievements of science and technology are highly visible. Its aim is not to replace these advances in knowledge, but rather to integrate them into the broader context of systems understanding, of appreciating sensitive relationships between people and their environments. Improvements in agriculture will only be genuine improvements when the values of whole systems of relationships are taken into account.

Action research integrates theory and practice in a manner that has a profound effect on the way we view the world. Its influences *how* and *what* one thinks, perceives, and decides. It says that how one knows (one's epistemology) is inseparable from what one knows just as it says that one's perceptions are inseparable from one's actions. Also, it is self-referential in that it includes how the researcher comes to know the system that he is a part of, and that he is endeavouring to change for the better. Finally, it gives direction to what this 'better' might mean.

From this perspective, the extension agent is not a mere transmitter of knowledge—a transferer of technology—but an action researcher. This professional, researches issues of relevance to his or her client farmers with the farmers themselves participating in the research and, in so doing, learning the ways of enquiry that characterise what is meant by science. This view, of course, does not negate the need for the conventional research scientist : indeed, the opposite is true. As farmers and members of their families become increasingly scientifically aware and literate, so they actively seek new technologies and information which scientists elsewhere will be generating.

The action researcher researches *enrich* farmers in a more profound sense than the increasingly common practice of research *on* farmers' fields (may be with the farmer as technician!). It is useful then to think of knowledge being generated through research *with* farmers, *for* farmers (conventional natural science) and *on* farmers (conventional social research).

Such a perspective has pervaded those in the Faculty of Agriculture and Rural Development at our University to the extent that it has become the dominant paradigm. Thus, we are able to increasingly illustrate that we are using systemic (systems oriented) action research to generate systemic action researcher !

Notes

1. Historically, the early Greeks at the time of Plato wanted to distinguish between mere 'opinion' and 'certainty'. The word episteme was used to refer to what was known to be true...to be certain knowledge. Epistemology then was the study of episteme and was the search for what was real (knowledge about how the world really is!). Episteme is the thing upon which we stand (*epi* means "upon", and *[hi]stanai* means "to stand"), and epistemology is the search for such a foundation. Also part of the search was the study of which methods best served this pursuit of true knowledge.

   Today, the link separating beliefs and opinions from certain knowledge is a very fuzzy one indeed. 'Established' knowledge remains open to future confirmation and correction. Science does not pretend to have passed the test of absolute certainty; statements of knowledge can be judged one against the other, and some of them can be accepted and used as the basis for action which others can be rejected. In a scientific climate such as this one, there is the phenomenon of alternative epistemological frameworks. Epistemologies can be explored and compared. They enable us to get insight into what are the underlying premises for anyone's thinking, decision-making and action. The adequacy of one's epistemology to match the perceived needs of a situation can likewise be evaluated.
2. Jurgen Habermas is a prominent thinker of the "Frankfurt School" of Social Theorists. His work is an attempt to unite theory and practice given the political and social context of the 20th-century. Habermas' critical social theory is informed by the belief that what we know and how we see things is determined by the social influences of the day. By reflecting on the way our knowledge is shaped by politics, technology and all the other sources of information, humans can challenge (critique) taken-for-granted assumptions. Rigorous criticism of existing modes of thought and action will emancipate people from a native acceptance of the 'real world' as the only possible world and thus enable people to search for new forms of 'reality' that are more rational, just and fulfilling.

References

Alverson, H. (1984). "The Wisdom of Tradition in the Development of Dryland Farming: Botswana", *J. Soc. App. Anthropology*, 43:1-8.

Anon, (1986). "Relevance of Education to Rural Development". Report of Regional Study Group Meeting, APEID/UNESCO Regional Office, Bangkok.

Auerswald, E. (1973). Families, change, and the ecological perspective. In A. Ferber, M. Mendolsohn, and A. Napier (Eds), *The Book of Family Therapy*. Boston, Houghton Mifflin.

Bartlett, P.G., and BROWN, P.J. (1985). "Agricultural Development and the Quality of Life: An Anthropological View". *Agriculture and Human Values*, 2:28-35.

BATESON, G. (1958). "Naven" 2nd Ed., Stanford Calif., Stanford University Press.

BATESON, G. (1971). "A Systems Approach". *Int. J. Psychiat.* 9:242-244.

BATESON, G. (1977). "The Thing of It Is", in M. Katz, W. Marsh and G. Thompson (Eds), *Exploration of Planetary Culture at the Lindisfarne Conferences: Earth's Answer*. New York, Harper and Row.

BATESON, G. (1979). "Mind and Nature: A Necessary Unity", Dutton, N.Y.

BAWDEN, R.J. (1988). "Models in Question—The American Land Grant System". An Essay in Sabhatical Reflection—Hawkesbury Agricultural College (AGPACK SS-107).

BAWDEN, R.J. (1989a). "Systems Agriculture: Learning to Deal with Complexity". MacMillan, N.Y.

BAWDEN, R.J. (1989b). "Towards Action Researching Systems". In proceedings of the Conference "Action Research in Education, Government and Industry". Brisbane, Queensland.

BAWDEN, R.J. and BUSCH, L. (1988). "Agricultural Universities for the 21st Century". A Paper prepared for US-AID.

BAWDEN, R.J., HANNIDAC, M. and GAHAN, C. (1989). "On Knowledge, Knowing and Change, Farmers in Semi-Arid, N.S.W. Report to NSW Department of Agriculture.

BAYLISS-SMITH, T.P. (1982). "The Ecology of Agricultural Systems". Cambridge University Press, Cambridge, 112p.

BERTALANFFY, L. von. (1968). "General Systems Theory", Braziller, N.Y.

BONNEN, J.T. (1983). "Historical Sources of U.S. Agricultural Productivity: Implications for R. and D. Policy and Social Science Research". *Amer. J. Agr. Econ.* 65:958-966.

BROKENSHA, D.W., WARREN, D.M. and WERNER, O (Eds). (1980) "Indigenous Knowledge Systems and Development". Landham, Md: University Press of America.

BUTTEL, F.H. (1985). "The Land Grant System: A Sociological Perspective on Value Conflicts and Ethical Issues". Agriculture and Human Values, 2.

CHAMALA, S. (1987). Strategies to Overcome Communication Deficiencies in Achieving Sustainable Farming Systems, in *Jnl. of Aust. Instit. of Agricultural Science,* 3:164-169.

CHAMBERS, R. (1983). "Rural Development: Putting the Last First". Longman, Burnt Mill, Marlow, Essex.

CONKLIN, H.C. (1957). "Hanumoo Agriculture: A Report on an Integral System of Shifting Cultivation in the Philippines". FAO Forestry Development Paper No. 12, FAO, Rome.

CONWAY. G.R. (1987). "The Properties of Agroecosystems". *Agricultural Systems,* 24:96-118.

COOMBES, P. and AHMED, M. (1974). "Attacking Rural Poverty: How Non-Formal Education Can Help". John Hopkins Univ. Press, Baltimore.

COSER, L.A. (1971). "Masters of Sociological Thought : Ideas on Historical and Social Content". Harcourt Brace Jovanovich Publishers N.Y.

CROUCH, B.R. and CHAMALA, S. (Eds) (1981). "Extension Education in Strategies for Planned Changed". John Wiley, Chichester.

DAVIES, A.J. (1988). "Communication in Agricultural Extension", *Jnl. of the Royal Agricultural Society of England, 149:* 155-164.

DORNER, P. (1983). "Technology and U.S. Agriculture, Technology and Social Change in Rural Areas", Summers, G. F. (Ed.), Westview Press, Boulder, Colorado.

DUNN, E.S. (1971). "The Nature of Social Learning" in "Economic and Social Developments: A Process of Social Learning". John Hopkins Press, Baltimore.

FALS BORDA, O. (1979). "Science and the Common People", cited in "Participatory Research Popular Knowledge and Power: A Personal Reflection", Hall, B.L. (1981), Convergence *14:* 6-17.

FLIEGEL, F.C. and VON ES, J.C. (1983). "The Diffusion - Adoption Process in Agriculture : Changes in Technology and Changing Paradigms" in "Technological and Social Change in Rural Areas". G.F. Summers (Ed.), pp. 13-28, Westview Press, Boulder, Colorado.

HABERMAS, J. (1972). "Knowledge and Human Interests" Trans. J.J. Shapiro, London, Heinemann.

HABERMAS, J. (1974). "Theory and Practice" Trans., J. Viertel, London, Heinemann.

HALL, B.L. (1981). "Participatory Research, Popular Knowledge and Power: A Personal Reflection". Convergence *14:* 6-17.

HANSEN, M., BUSCH, L., BURKHARDT, J., LACY, W.B. and LACY, L.R. (1986). "Plant Breeding and Biotechnology". Bioscience 36:29-39.

JACKSON, D., and SATIR, V. (1973). "A Review of Psychiatric Developments in Family Diagnosis and Family Therapy", in D. Jackson (Ed.)., *Therapy, Communication and Change.* Palo Alto, Science and Behaviour Books.

JAIN, H.K. (1988). "Role of Research in Transforming Traditional Agriculture: An Emerging Perspective". ISNAR Reprint Series No. 24.

KELLY, G. (1955). "The Psychology of Personal Constructs". Vols. I and II. W.W. Norton, N.Y.

KEMMIS, S. *et. al.,* (Eds) (1982). "The Action Research Reader" Geelong, Victoria, Deakin University Press.

KEMMIS, S. (1985). Action Research and the Politics of Reflection. In *Reflection: Turning Experience into Learning,* D. Boud, R. Keogh and D. Walker (Eds), London, Kogan Page.

KENNEY, B. (1979). "Ecosystemic Epistemology: An Alternative Paradigm for Diagnosis". *Family Process,* 18(2): 117-129.

KLOPPENBURG, J. and KENNEY, M. (1984). "Biotechnology, Seeds and the Restructuring of Agriculture", Insurgent Sociologist, *12:* 3-17.

KORTEN, O. and KLAUSS, R. (Eds.) (1984). "People Centred Development". Kumarian Press.

KINGMA, O.T. (1985). "Dilemmas in Productivity Growth and Development of Agriculture"; 29th Annual Conference of the Aust., Agric. Econ. Society.

LANGER, E. (1985). "Mindlessness/Mindfulness". Unpublished MS. Harvard University.

LEWIN, K. (1946). "Action Research and Minority Problems", *Journal of Social Issues, 2:* 34-46.

LOVELACE, G.W. (1984). Cultural beliefs and management of agroecosystems. In A.T. Rambo and P.E. Sajise (Eds). "An Introduction to Human Ecology Research on Agricultural Systems in Southeast Asia". pp. 194-205. East-west Environment and Policy Institute and University of the Philippines at Los Banos.

MACADAM, R., and BAWDEN, R.J. (1985). "Challenge and Response: Developing a System for Educating More Effective Agriculturalists". Prometheus, *3*:125-137.

MEDEWAR, P. (1979). "Advice to a New Scientist". Harper and Row, London.

MELLOR, J.W. (1986). "Agriculture on the Road to Industrialisation" in "Development Strategies Reconsidered". J. Levin and V.Kallab (Eds). US Third World Policy Perspective No. 5 (Transaction Books for Overseas Development Council).

OAKLEY, P. (1968). "Extension and Technological Transfer: The Need for an Alternative Hort. Science", *23*:482-485.

PENMAN, R. (1987). "Communication Re-constructed". Paper presented at the Australian Communication Association Annual Conference.

RAMBO, A.T. (1985). "Applied Human Ecology Research on Agricultural Systems in Southeast Asia". Paper presented to Workshop on Agricultural Systems Education, The University of Hawaii College of Tropical Agriculture and Human Resources, 15-22 July, 1985.

REASON, P., and ROWAN, J. (1981). "Human Inquiry: A Source Book of New Paradigm Research", John Wiley, N.Y.

RHOADES, R.E. (198/). "Farmers and Experimentation". Discussion Paper 21 Overseas Development Institute, Ag. Administration Network (1983).

RHOADES, R.E. and BABBINGTON, A. (1988). "Farmers Who Experiment: An Untapped Resource for Agricultural Research and Development". Proc. International Congress on Plant Physiology, New Delhi, India, (in press).

ROGERS, E.M. and SHOEMAKER, F.F. (1971). "Communication of Innovations: A Cross-Cultural Approach", McMillan, London.

ROLING, N. (1985). "Extension and the Development of Human Resources: The Other Tradition in Extension Education". Paper AERDC Conference. Reading University, Reading.

ROLING, N. and de ZEEUM, H. (1983). "Improving the Quality of Rural Poverty Alleviations". Int. Agr. Centre, Wageningen, Holland.

SALMON, P.W. (1980). Report to Joint Advisory Committee on Agricultural Extension: Strathfieldsaye "Think Tank" October 8-10. Monograph, School of Agriculture and Forestry, University of Melbourne.

SELVINI-PALAZZOLI, M., CECCHIN, G., PRATA, G., and BOSCOLO, L. "Paradox and Counterparadox" New York, Jasŏn Aronson, 1978.

SMITH, K. (1983). "A role for community psychologist: As participant-conceptualiser". *Australian Psychologist*, 18(2): pp. 143-159.

SPENCER, H. (1891). "The Study of Sociology". Appleton, N.Y.

SUSMAN, G.I., and EVERED, R.D. (1978). "An Assessment of the Scientific Merits of Action Research", *Administrative Science Quarterly*, 23. pp. 582-603.

SWAMI RANGANATHANANDA (1985). "Science and Human Values". Inaugural Lecture of 9th Foundation Week, National Academy of Agricultural Research Management, Hyderabad, India.

WOLF, E.R. (1982). "Europe and the People Without History". University of California Press, Berkeley, 503p.

2

The Role of Development Communication in Agriculture: Status and Trends with Special Reference to Language

Md. Mahfuzul Huque

Introduction

Technological innovations, based on research tradition of various fields in the scientific arena around the globe, have brought about revolutionary changes to analyse, understand, and solve human problems of varied nature and complexities. Of specific mention, technological development in agriculture and its diffusion to farmers field have changed the method of farming—the oldest profession of any place, race, and culture. These developments have increased quantitative and qualitative production, decreased cost per unit of area and less use of human labour and draft power. The consequence is that men and machines are at constant war for work and continuous victory has been for the latter.

That is a situation one can see in the advanced countries of the West. By contrast, faced with many socio-economic hard realities, most Third World countries cannot produce enough food to eat, adequate fibre to clothe, required employment to earn livelihood and thus they remain unable to have a decent living. These countries mostly include those of Latin America, Africa and Asia. One of the many characteristics of their economy is typified by traditional agriculture where agricultural growth generally remains low and unchangeable over time. The United

Nations Survey of 1976 on 100 developing countries (with a total population of 1.8 billion) showed that between 1970-75, only 27 per cent of these countries reached or exceeded the target growth rate for agricultural sector of 4 per cent a year. Only 10 per cent of the combined population lived in these countries. Between 1971-75, per capita agricultural production in 87 countries (with a total population of 1.75 billion) was less than 2.5 per cent a year. This was less than the average annual rate of population growth during the same period (Ball, 1981). This corroborates to an earlier USDA production in 1964 that total world population would increase more rapidly than total food production (Rogers and Svenning, 1969).

Development planners and policy makers of these countries have been critical of the research and extension systems for this situation. However, with gradual establishment of 13 international agricultural research centres, especially from the early 60's and strengthening the national research system in many countries, a stream of productive technologies started emerging out. The social scientists side by side came up with many development models, the first of its kind was probably the RD & D propounded by Brickel in 1961 and later by Clark and Guba (1965). Since then hundreds of such models have been postulated, a compilation of these appear in the works of Havelock *et al.*, (1975). These were built around a given frame of reference and therefore each bears some merits and demerits on a policy adoption level. However, prototype of the RD & D model seems to exist in agriculture and industry.

The Nature of Technological Information Explosion

Scientific processes usually begins on recognition of a problem that seems affect the lives of plants, animals and humans. These often times lead to the discoveries of appropriate technologies for survival of human race and its civilization. Depending on the nature and complexity of a problem, the search for technology as a remedy may take months to several decades. None knows how many scientists shall have to work and for how long before reaching to a solution for the AIDS—the most dreadful disease that appeared since the early 80's.

Research activities with all its costly endowments have been increasing progressively in all fields of specialization. The knowledge system thus enriched gets doubled every seven years (Morris and Elkins, 1978). Libraries have been the most important repository where information are preserved in the traditional form of books, periodicals, annuals, journals, reports, proceedings, thesis and dissertations and in modern forms—microfilms, slides, tapes, records and most recently in the computerized information—retrieval systems.

The nature of information explosion could be realized by the ever increasing number of scientific journals and periodicals published over the past two centuries. The number of publications was only 100 at the beginning of the 19th century, reached to 1,000 in 1850, reached further to more than 10,000 in 1900 and finally approached to 1,00,000 in 1960 and should this rate continues—it would be nearly two millions by the end of this century. Hillard, an educational broadcast scientist even predicted a higher rate of "information explosion". He held that by the time a child is born and the day he comes a college graduate, the amount of knowledge in the world becomes four times and later 32 times as great when he reaches at his fifty. About books, during mid-sixties, the output on world scale approached the prodigious figures of 1,000 titles per day. On a world-wide level, scientific and technical literature stands at a rate of 60,00,000 pages a year. The U.S. Government Printing Office sells about 80 million publications yearly (Viloria, 1975).

The status of agricultural literature published is also quite high. An annual output within the range of 2,10,000 to 2,50,000 entries is estimated (AGRIS, 1973). With regard to publications of rice alone, the most important cereal crop in Asian countries, the number of journal articles published in 1970 went up by two-folds in 1977 (Busch, 1981).

This is the nature of information explosion which poses a serious problem for the librarians, documentalists and development communication specialists as to how to move the huge bulk of information to the consumers—men working in teaching, research and extension profession in agriculture. The fact that so much information are generated, so quickly, in so many fields and for too many consumers has made information retrieval system a difficult task.

Most research in the past was done in isolation keeping little in mind about the end users. This sporadic attempt led to only one invention in a thousand surviving and being accepted by the public (Linton, 1936). The situation has not improved even today. Despite the appearance of so many communication channels which Schramm calls "big media" and "little media", 90 of every 100 inventions, which took decades of work, get lost (Lionberger and Gwin, 1982). In the agricultural technology diffusion process, a chronic "information crisis" can be located at all levels—the research system, linking system and most seriously the client system. The nature and intensity of the factors responsible for this information crisis may vary from one level to another. However, it is quite apparent that the role of mass media progressively diminishes along the vertically downward communication at the one-to-one level of the research-linking-client systems.

Role of Communication Channels in Agricultural Technology Transfer

Communication technologies must not only move the needed amount of information to the clients, but also the flow should be as fast as possible. We recall that Napolean was finally defeated by the allied forces in the battle of Waterloo on 18th June, 1815. But it took complete three days reach this news of historic event to the British Government which was located only 200 miles away from the war field. Further, it took 44 days after the battle before the news was published in the New York newspapers (Barry *et al.*, 1965). Today with the tremendous growth of modern communication technology, people can know events that happen in any part of the world in almost the same moment of their occurrence. With increased importance of communication and growth of communication technology, line item in the budget of industrial and agricultural organizations needs to be proportionately allocated. *The Business Week* predicted, for example, that by 2029, "no industry will be bigger or reach further into the lives of Americans than communications". New communication technology may make it possible for people all across the world to see each other in colour as they talk. Conducting international group meetings might one day be possible without leaving one's office in his country (Mitchell, 1982).

In agricultural development, the role of technological information for increased farm production has been well established by thousands empirical evidence since Ryan and Gross studied the diffusion of hybrid seed corn in 1942. Two systems, the linking and client, must have adequate access to information through both interpersonal and mass media. Presented below in brief the role of development communication in the diffusion of farm technology on a cross-cultural foundation.

In a study conducted a decade ago with Lewis County dairy farmers it was found that extension agent dominated in two of the five attributes e.g., 'most helpful' and 'most trustworthy'; but magazine dominated in two other attributes—"most often used for technical information" and 'most likely to have latest agricultural developments'. Respondents were asked if magazines, newspapers, radio and television were: (1) the most important and (2) the most convenient source for agricultural information. For item one and two, magazines were rated by 57 and 53 per cent of the farmers, respectively; newspapers by 28 and 17 per cent; and radio by 13 and 26 per cent, in that order of the items. Interestingly, television was not perceived as significant medium in either case (Awa and Crowder, 1978) although some 25 years ago, one study showed that 92 per cent American farmers in a County possessed television sets, 90 per cent used to watch television and 25 per cent watched regularly the specific programme "Better Family Living" (Knox, 1962).

Evidence is clear that the information seeking behaviour of the American people including the farmers undergoing a drastic change. By 1985, 40 per cent U.S. television households subscribed to cable television and it is expected that 70 per cent of American households would adopt it by 2000. Over the past decade, 20 per cent of households accepted video cassette recorders, 15 per cent have at least a micro-computer. It is estimated that 54 per cent of the American work force are information workers (Rogers, 1986). The new communication technologies are making their way into the farming communities of the West. For example, 200 Kentucky farmers received Green Thumb videotext system under a joint project of Federal Extension Service of USDA and Extension Service of the University of Kentucky. In this system farm families have access to 900 frames of information, each in a coloured slide dealing with farm markets, weather report, agricultural and home economics, occupational information for youth and news of local farmers' meeting (Rogers, 1986). Also diffusion of these technologies have been reported in most European countries e.g., Netherlands, Belgium and Canada at a rate comparable to that of the U.S.A. It is only Japan in Asia that could be considered to have entered into the "information society'" of Western nature. Diffusion of new communication technologies by the farmers has been today a new area of research.

By contrast, innovation diffusion in most developing countries has remained heavily dependent on the interpersonal media over a long past. Communication studies in farm technology diffusion in India showed that both radio and television are effective, however, television is more effective than radio (Dey, 1968). The Poona radio forum experiment conducted in the late fifties showed a significant knowledge gain by participants of the forum over those radio listeners but without forum (Rogers, Brown and Vermilion, 1977). In Bangladesh, 35 listeners' clubs were organized in 1977 in a joint project between Radio Bangladesh and Dhaka based UNICEF. In the beginning activities were regular, but later these were deteriorated because of lack of well trained staff (UNESCO, 1981).

The role of educational television was experimented with farmers around Delhi in 1967. In the Pilot Rural Agricultural Television Project, called *Krishi Darsan*, about 100 sets were installed for community viewing and two programmes on agriculture were telecast in a week. The project was later expanded to Haryana and Uttar Pradesh and the concerned governments became partners in the experiment. Evaluation conducted by National Council for Educational Research and Training (NCERT) conclusively showed that the programme had contributed significantly to attitude changes and knowledge gains which finally led farmers adopt new technologies (Hancock, 1984).

Indian Farmers' perception of credibility of sources in typical rural settings are exclusively towards the interpersonal ones (e.g., demonstration, scientists, extension agent, progressive farmers etc.) than that of mass media. However, in progressive villages and with some literacy rates of farmers, credible sources include a combination of both inter-personal and mass media—radio, television and print (Singh, 1971; Sandhu, 1970; and Singh, 1981).

In the absence of a common methodological design, generalization may be unwarranted. However, several earlier studies in Bangladesh seem to suggest that about 70-75 per cent of farm information reach the rural mass through interpersonal sources and the rest through mass media. Members of the peer group, extension agents and group meetings are the most important sources; while among the mass media, radio farm broadcast is the most important channel followed by extension print media. But newspaper, farm magazine and television have insignificant role in technology transfer (Huque, 1971; Rahman, 1974; Latif, 1974). However, an insignificant increase in the use of mass media by the farmers seems noticeable over a period of past two decades (Bhuiyan, 1988).

Generally in Latin American countries, change agents perceive books and technical publications as most useful, but their perception of newspapers and magazines was second in place, however, well ahead of the electronic media—radio, television and cinema. With major thrust in entertainment, these media were assigned less credit for development communication (Deutschman *et al.*, 1968).

The limited use of mass media by the farmers in developing countries has often been blamed for large scale illiteracy and less ownership to electronic media. While in the Philippines, with about 90 per cent literacy rate, higher per capita income, and more ownership to radio and television than either of Bangladesh and India, farming communities show a similar trend—a heavy dependence on interpersonal sources for farm information. Based on an extensive review of contemporary development communication researches, Jamias' generalization show that interpersonal sources—e.g., friends, neighbours, extension workers and commercial dealers are the major sources, while except radio, mass media play a minimal role (Jamias, 1979). In a study with Masagana—99 farmers around a radius of 2-3 km off the IRRI campus, 62 per cent of the credibility scores was assigned to five interpersonal media and the rest 38 per cent to five mass media. Level of perception of effectiveness of the mass media was confined from poor to moderate; but that of interpersonal media was from moderate to high. In general, a positive significant relationship was observed between perception of a medium by the farmers and its corresponding use by them (Huque, 1982).

The production and use of print media in developing countries also show a similar picture. Eighteen selected Asian countries, with 28 per cent world population, produced book titles slightly over than 7 per cent of world's total production. This represented only 32 book titles per million population compared to a world average of 127 or an average for Europe of 418 (UNESCO, 1965). The volume of development communication print media generally released by the extension agencies in this region seems to be very inadequate compared to their demands. The distribution mechanism is also inappropriate. In the Philippines, it is said that many primers, pamphlets, bulletins and training materials for extensionist are produced by educational and research centres, but they do not reach the extension workers (Bernardo, 1981).

Language and Communication in Cross-Cultural Communities

Earlier we saw how developing countries remain in chronic technological information crisis due mainly to heavy dependence on interpersonal channels in their communication policies. On an average,

70 per cent of the working hours of human beings is spent in communication (Allen, 1982). Research studies, however, show that up to 50 per cent of the information transmitted is interpreted incorrectly (Mitchell, 1982). In terms of extension work Crouch and Chamala (1981), referring to Potter *et al.*, (1974), stated that in extension work, about 47 per cent of the time of the extension officers was being spent directly and indirectly communicating with farmers. This study seems to have been conducted in an advanced country, possibly in Australia. Since information sharing in developing countries mainly takes place via interpersonal channels, the percentage of time devoted by grassroots level extension workers would be much higher than Potters' estimation especially in the Training and Visit System. And if the 50 per cent death or distortion of message occurs, then the 40 developing countries where T & V have been introduced, have been paying every year a significant portion of their organizational budget without any return. Language problems, among others, have been a significant factor for many of these.

Distribution of Major Languages of the World

It is uncertain how many languages exist in the world. Estimates vary from 3,000 (Encyclopedia Britanica, 1971) to 10,000 (Barry, 1965). Many of these disappeared from the world without trace and almost all are in a constant change. Fortunately from the viewpoint of technology transfer, only 12 of these languages are spoken by more than 75 per cent of mankind (Escarpit, 1966). The distribution of these languages is presented in Table 1.

Languages are unevenly distributed all over the world by families and sub-families. However, the Indo-European language family has shown maximum spread which includes 9 of the 12 most widely spoken languages in the world. These include English, Russian, Spanish, Hindi-Urdu, German, Portugese, Bengali, French and Italian. The family tree of the Indo-European Languages shows that Portugese, Italian, French and Spanish originate from one branch, the Latin. Similarly, English and German with two more stem from another branch, while Russian along with seven more originate from another branch. In the politics of Persian Gulf and South Asia, there are four languages that originate from one branch. These are Hindi-Urdu, Bengali, Marathi, Pushtu and Persian (James and Davis, 1983).

Table 1: **Distribution of Major Languages of the World**

Language	*Country in which spoken by largest group*	*(% of population)*	*Population (In 000's)*	*Literacy rate (%of population*	*Per capita income (U.S. Dollar)*
1	*2*	*3*	*4*	*5*	*6*
Mandarin Chinese	China (Mainland)	81	10,45,537	75	566
English	Barbados	98	253	99	3,040
	Jamaica	98	2,288	73	1,340
	United Kingdom	98	56,458	99	7,216
	Trinidad	97	1,204	86	6,800
	Australia	92	15,763	100	10,282
	United States	86	2,40,856	99	13,451
	Canada	58	25,625	99	13,000
	Guyana	45	771	86	457
	New Zealand	91	3,305	99	7,282
Russian	USSR	59	2,79,904	99	2,600
Spanish	Cuba	98	10,221	96	1,590
	Dominican Republic	98	6,785	77	1,221
	Colombia	97	29,956	80	1,112
	Costa Rica	97	2,714	90	1,352
	Nicaragua	96	3,342	66	868
	Uruguay	94	2,947	96	1,665
	Venezuela	94	17,791	88	2,629
	Honduras	93	4,648	56	815
	Chile	92	12,261	90	1,950
	Panama	92	2,227	87	1,970
	Ecuador	86	9,647	90	1,299
	El Salvador	86	5,105	50	858
	Argentina	84	3,186	94	2,331
	Mexico	81	8,170	74	2,082
	Spain	73	39,074	97	4,490
	Guetamala	55	8,600	48	1,000

Contd.

1	*2*	*3*	*4*	*5*	*6*
Hindi	India	42	7,83,940	36	150
Urdu	Guyana	45	771	86	457
German	East Germany[1]	100	16,692	99	8,000
	West Germany[1]	100	60,734	99	9,450
	Austria	99	7,546	98	8,280
	Luxembourg	93	367	100	10,444
	Switzerland	69	9,466	99	14,408
Arabic	United Arab Republic	98	1,326	56	23,000
	Jordan	98	2,756	31	—
	Yemen (North)[2]	98	6,339	20	475
	Yemen (South)[2]	98	2,275	39	316
	Saudi Arabia	96	11,519	50	11,500
	Tunisia	93	7,424	46	1,163
	Lebanon	92	2,674	75	1,150
	Oman	89	1,271	20	—
	Libya	88	3,876	60	7,000
	Algeria	87	22,817	52	2,085
	Syria	87	10,931	78	—
	Kuwait	85	1,771	71	11,431
	Mauritania	82	1,691	17	466
	Iraq	92	16,019	70	2,410
	Morocco	65	23,667	70	630
	Sudan	48	22,932	20	361
	Chad	33	5,231	17	88
Portugese	Brazil	98	143,277	76	1,523
	Portugal	100	10,095	80	1,930
Bangali	Bangladesh[3]	99	104,204	25	113
Japanese	Japan	99	121,402	99	10,226
French	Haiti	99	787	23	300
	France	86	55,239	99	10,260
Italian	Italy	99	57,226	99	6,447

Source: (a) Column 1-2, adopted from Rustow (1968).
(b) Column 3-5, adopted from the World Almanac and Book of Facts (1988).

1 Shown by Rustow as one Germany.
2 Also shown as One Yemen. Separated in both cases based on World Almanac's data for columns 3-5, however, data in column 2 correspond to those of Rustow.
3 Computed based on population statistics of the aboriginals who sharply differ in linguistic characteristics in Bangladesh.

Fishman (1968) drew contrasts between linguistically homogeneous and heterogeneous using data from various sources. Linguistically homogeneous, according to him, means a polity in which a single language is natively spoken by at least 85 per cent of the population and in which no significant minority is present. By contrast, linguistically heterogeneous politics are those in which a significant minority is present among the rest 15 per cent of population and in those no single language is natively spoken by 85 per cent or more of the population. Based on this, his analyses bear significance to the understanding of linguistic phenomena relevant to development communication. He observed among others, that linguistic homogeneous politics generally: (a) are smaller in size, (b) have a higher population density, (c) have a lower proportion of agricultural population, (d) are highly urbanized, (e) have at least a medium per capita GNP, (f) have at least stable economic growth, and (h) tends to be religiously homogeneous. Conversely, linguistically heterogeneous politics tend to be in the opposite direction.

Language Diversity as a Problem in Global Communication

Less than one-third of the countries of the world are the major producers of technology and hence they are the custodians of research-based knowledge (Ahmed, 1987). These are mostly the advanced countries of the West, particularly those of the European and the Americas. It is estimated that 80 per cent of scientific works are published in only five languages—English, French, German, Japanese and Russian (Watson, 1975). Of these, Russian and Japanese have been the exclusive language of the USSR and Japan, respectively. The rest three—English, French and German are the major languages used in scientific publications among some of the countries of Europe and the Americas. But interestingly, there are few countries in this region which have satisfactory access to all these three languages. However, barrier to free flow of technological information due to language diversity can be taken-care of by translating the materials. But considered globally language remains a major obstacle in sharing technological information among the nations of the West and the East. Only materials published in English can reach to the Eastern nations satisfactorily than those published in the other two major languages namely, French and German. On an over all consideration, 50 per cent of the world's scientific literature is published in languages that most scientists can not read (UNESCO 1957).

Characteristics of Language and their Role in Development Communication

The effect of language characteristics in development communication can't fully be realised in isolation, rather the issue should be addressed on a broader perspectives. Analyses, therefore, would encompass the various characteristics of selected languages that affect human communication and correlate phenomena in development communication with special emphasis to agriculture. These would also embrace the linguistic features of the people who belong to the linking and client systems.

A common characteristic of languages is that these, by and large, represent a dynamic system. Changes occur in orthography, vocabulary, word form, sentence, structure, syntax and in phonetics. Even a language which is widely spoken throughout a country shows variations of many form although such varieties may not lead to serious unintelligibility among the people living in different parts of that country. For example, although Chinese are largely homogeneous, the language of Canton region is somewhat different from rest of the country (Deutsch, 1966).

The nature and extent of variations seem correlate with the geographical area which may be confined within the national boundary or exceed beyond that territorial limit. English, for example, used by greatest number of world's scientists, is spatially dispersed over the globe. But variations can be identified in English of the British, Americans, Canadians, Australians, South Africans, Filipines and, of course, the Indian subcontinent. Clatthorn (1971) prefers to call these as the dialects of English. In the American colonies and in England, the word *druggist* used to mean a retailer of drugs. By 1850, the English substituted the word by *chemist*, but the Americans still use it as druggist. What an American calls today *appartment*, a British calls it *flat* and the latter began it so from the early eighteenth century. *Truck* and *elevator* of the Americans means *lorry* and *lift*, respectively to the English (Glatthorn, 1971).

The acquired foreign language used by the people of a country, whose origin could spatially be located far away from that country, shows much variations. Historically many Asian and African countries were ruled for long by the colonial powers, especially of the West. People of these colonial had learned the languages of their colonial powers and they maintained the characteristics once acquired unchanged although these foreign powers had quit long ago. Thus,

English of the Filopines, called Filippine English, remains patterned after the Americans and that of the Indian sub-continent after the British. Any one from this sub-continent on a visit to the Philippines will soon discover that his 'groundnut' has been changed overnight to 'peanut', 'maize' to 'corn' and that a vegetable vendor will stare at him when he seeks 'brinjal' for 'eggplant'.

Language differ by their way in which sentences must reflect the world of experience and by the grammatical way in which sentences must be constructed. For examples, differences of this nature can be noticed in asking a person what his name is in the following four languages (Bloomfield, 1974):

English	*French*	*Italian*	*Russian*
What is your name?	How you call you? (*Comment vous appelez-vous?*)	How self calls? (*Come si chiama?*)	How you they call? (*Kak vas Zevut?*)

We noted earlier that Arabic is one of the World's major languages spoken by largest group of population in 18 countries. It is also used by millions of Muslims around the globe at least five times a day for religious purpose. But it is distinctively so different in its many linguistic characteristics that effective communication can hardly occur. A message encoded in classical form and transmitted through complicated channel could be decoded by only 3-6 per cent satisfactorily and by 8-10 per cent partially, notwithstanding of the fact that Arabic has enriched many other languages by hundreds vocabularies and terms (Matter, 1975). Thus, Deutsch (1966) correctly says: Language communities are bounded by barriers of dialect and language, geographically between regions, sociologically between strata, and historically in time.

Some of the problems in Arabic arise from semantic pluralism. Laxicographers classified between 300-400 words which refer either of two opposite concepts e.g., distance and proximity, visible and hidden and so on. Antonyms of this nature is found in Tausug dialect of the Philippines, for example, *habagayan* means both friend and enemy (Spadley, 1979). The word is not technical at all. But it has serious implication in development communication in constructing a message on biological control of insect pests that appeals to farmers: "Don't kill your *friend*, Ladybird beetle, because it kills your *enemy*, Green leaf hopper".

Communication becomes hard in a language that possesses large number of synonyms of semantically equivalence. Referring to Yazigi, Matter (1975) reports that there are 80 synonyms for honey, 200 for bird, 500 for lion, 1,000 for sword, and 4,000 for misfortune. Camel is an indispensible companion of the Arabs as bullock is to the South Asian farmers. Arabic provides 1,000 names for that animal which denote its numerous breeds at different stages of growth (Hitti, 1961). An Iraqi scholar with thorough familiarity of the Arab World identified 17 different words for camel-dung, each being the only natural one for family and village use in its own area (Smith, 1983).

For technology transfer, a common problem in most languages is a serious lack of adequate vocabularies, especially in agricultural and technical terms. Arabic is very much expressive by its exceedingly high number of synonymous terms. Contrary to this situation, encoding development messages in classical Arabic becomes hard because many concepts have no equivalent terms in the language. It is surprising, however, that between eighth and sixteenth centuries, it was a world-wide vehicle for scientific studies, comparable in importance only with Latin (Mattar, 1975).

This problem exists also in the Philippines languages. In the absence of adequate vocabularies, a common term includes more than one insect pest e.g., *Waya-waya* stands for six different insect pests in Ilonge speech community (Litsinger *et al.*, 1980). In Cagayan Valley, few farmers (7 per cent) could distinguish rice greenhorned caterpillar from rice skipper larvae even though both are large with conspicuous morphological differences. (Litsinger *et al.*, 1982). One of the reasons for their inability could be attributed to their unfamiliarity to the local terms which the community has one for each. In Pangasinan (Ilocane speech community), no distinction is made between any of the stemborers. Rice caseworm has a local term in Ilongo dialect, but it has not in Ilocano dialect. Further, in Ilocano, Ilongo and Tagalog dialects, there is a general term for hoppers but no distinction is made, for example, between green leafhopper and brown plathopper. Local names for rice bug translate as 'bad small' which can also mean beetle (Litsinger *et al.*, 1980).

The Bengali language also suffers from adequate agricultural vocabularies and native terms vary from one area to another. For example, jute hairy cater-piller is variously known as *Bich poka* and *Sua poka;* and jute field cricket as *Maather jhee jhee poka, Urechunga* and *Ghogra.* The two species of jute mite are called by a common term

Makar adding a colour term *saada* (white) and *laal* (red) as suffixes to it for the first and second, respectively (Ahad *et al.*, 1987). Also in Bengali different crops are variously known in different speech communities. For example, sugar cane has at least four local terms, guava has five, papaya three and tomato three.

In the absence of direct vocabularies for some terms, circumlocution has been followed in some languages. The process is filling the gap by employing several words or awkward phrase. In Bengali, for example, two sugar cane insects whose scientific names are: *Chile infuscate* llus and *Chile auricilius*, have been termed in the literature as *Akher doger dora kata mazra, poka,* and *Akher gater mazra poka*, respectively, (Department of Agricultural Extension, 1988). It is doubtful if such localization of terms leads to effective communication, or create noise' in communicating messages with the illiterate farmers. A severe lack of local terms, especially in science and technology, have also been reported in African languages (Doob, 1961).

One of the interesting areas of vocabulary deficiency lie in colour spectrum. In Tagalog, local terms used are for white, black, red and yellow. Spanish lean words such as *berde, asul, chokolate and rosa* have been put in use for green, blue, brown and pink, respectively. But direct expressive local terms are lacking for gray, beige, tangerine, purple and cream (Pascasio, 1964). African languages, for example, Gunda and Zulu, have the colour spectrum at different and fewer points. Empirical data from Africa show that respondents who new English could respond to colour more readily and accurately than those who were only exposed to their local language which are deficient in colour terms (Doob, 1961). Doob, however, does not recognise lack of colour terms to affect communication. But colour identification has significant bearing in agricultural research to the scientists, in extension work to the change agents, and in accepting technology to the client system. For the latter two cases, a fair perception of colours would be helpful in communicating messages on the morphology of plants and animals, to identify and differentiate among similar farm inputs e.g. fertilizers and insecticides.

Units of measurements are complicated because these are not consistent on a wide geographical divides. At present there are three systems in Bangladesh, indigenous—the traditional ones inherited from culture, the British system—introduced possibly by the British from a long past, and the metric system introduced in recent days. Generally farmers have been well accustomed to the traditional system. However, variations in the traditional system occur even in the same term from one locality to another. In farm land measurement, three *bighas* equal to one acre. But the area under bigha varies widely throughout the country. Common units of weights in the traditional system are *maund*, *seer* and *chatak*. But maund varies widely from one area to another. In recent years, the units of measurement used in communicating the impact points of the T & V system are of metric system which most farmers fail to comprehend accurately (Karim, 1987).

Linguistic Characteristics of the Linking and Client System

Foregoing discussions imply that linguistic characteristics could be both barriers and facilitators in development communication. These characteristics may affect the communication skills of the people that belong to any of the three systems—research, linking and client. We will discuss specifically two important issues—what are some of the linguistic characteristics of both the linking and client systems? Second, how the two systems converge or diverge in sharing meaning with which they oftenly engage ?

Linguistic characteristics of the linking system seem to be more complex in a multilingual community. However, change agents in three speech communities (Iliocano, Ibang, and Itawes) of the Philippines showed good proficiency in Ilocano, but they had various levels of proficiency in Ibang and Itawes. Further, they were generally fluent in English (Juliano, 1981). Employed in Leyte and Southern Leyte provinces of the same speech community (Cebuano), 56 per cent of the agents know five languages and 34 per cent knew four languages. Further analysis show that Leyte workers knew significantly more languages than Southern Leyte workers. Considered generally and combining the data of both groups, subjects possessed highest proficiency in Cebuano followed in descending order by English, Philippine (national language) and Waray (a language of a nearby speech community). However, proficiency level for English and Philippine were almost at equal strength (Huque, 1986). Data on extension workers' attitude toward Cebuano (the major language of the area where they are employed) reveal interesting finding. Only 48 per cent of the Leyte technicians had highly favourable attitude, but this per cent for Southern Leyte technicians went up to 86. Thus, extension workers by group differ significantly in their attitude toward Cebuano, the main language of the clients with which agents have to conduct development communication campaigns (Huque, 1986). Place of posting of the agents within the geographical area of one's birth place varied significantly. While 80 per cent of the agents in Leyte were born in that province, while only 50 per cent of the agents in Southern Leyte were born in that province. Thus many of the 50 per cent of the agents in the latter province who came from a different province (as their birth place) might have a different mother tongue and hence possessed poor proficiency level in Cebuano. In Chidalo community of Ethiopia, lack of knowledge in local dialects has been a major obstacle in establishing a dialogue with rural people (Nekby, 1971).

The greatest obstacle in communication can be attributed to the unfamiliarity of the local terms by the agents in agricultural technology transfer. Majority of the agents employed in a Tagalog community were not familiar with more of the 731 key rice terms used in Philippines extension publication (Joaquin, 1978).

Clients generally show a poor proficiency in their own national language. The official language of Sudan is Arabic. In a survey with a Western District of Darfur province, it was revealed that only 10 per cent of the respondents had "good" knowledge of Arabic and 30 per cent had some. Half of the men and hardly any of them could read or write, and younger people know more Arabic than older people did (Jernudd, 1968).

In a multilingual community, the pattern of use of languages by the clients posses a complex phenomenon. The nationwide survey with different groups of people, it was revealed that local dialects were used most (47 per cent of the respondents) followed by Philippine (National language) and English equally (23 per cent of the respondents in each case). Philippine was effectively used in agrarian reform and green revolution, but English was effectively used in family planning, public health and crop production. Analysis, however, was not addressed to ascertain the relative effectiveness of the three languages—a local, a national and a foreign in communication.

Linguistic diversity between extension agents and their clients seems to affect adversely communication goal in African countries. For, extension staff members have the temptation to pepper their instruction and explanation with unfamiliar jargon. Farmers, on the other hand, have their peculiar phrases and metaphors which are often equally incomprehensible to the field extension staff (Lily, 1975). Several studies conducted in India tend to suggest that language incompatibility exists between extension workers and clients. This situation also exists in the appropriateness of word selection and interpretation which brings obstacle to mutual comprehension. The communicators develop a language jargon of their own which goes beyond the comprehension of an average farmer (Singh, 1981).

Co-publication: A Strategy to Bridge the Language Gap

The foregoing discussion demonstrate well how language can contribute to the effectiveness of communication. To make every communication meaningful, communication specialist must manipulate all variables in the basic elements (e.g., S-M-C-R) in the best possible order along with treatment of message in the best available symbol system. Development communication specialist, therefore, must understand :

Who (scientists, extension agents, local leaders, farmers etc.) uses.
What language (pure, classical, colloquical, local, foreign etc.) for
What messages (research oriented, theory based, developmental etc.) through
What channel (journal, report, demonstration, radio etc.) for
Whom (scientists, science writers, agents, lay leaders, farmers etc.) and for
What purpose (inform, teach, direct, please, persuade etc.).

The language barrier to farm technology transfer was first globally recognized in 1979. Information scientists from all over the world met in an IRRI sponsored workshop and adopted a policy of translating the works of the international research centres in various languages other than those used by the centres. The workshop observed: "Translation is an acute problem with no easy solution. The centres must come to grips with the problems of translation".

There are 13 international agricultural research centres coordinated by The Consultative Group on International Agricultural Research (CGIAR). Each centre publishes its works in many form only in selected languages, but often times in its selected lone language, for example, the International Rice Research Institute uses English, International Institute of Tropical Agriculture (IITA) uses English and French. However, English seems to be the most widely used language in the publications of international research centres. But the reality is that knowledge of English progressively drops with men employed in the research and linking systems and finally leads almost to nil in the client system—which constitutes the largest group of research based information consumers. Ironically, hardly one per cent of the farmers in the developing countries could comprehend the message from English publications.

To move the farm production messages to the linking and client systems, publications must reach these consumers group in the latters' own or known languages. This refers to IRRI's co-publication strategy, a co-operative endeavour whereby the original publisher (IRRI) grants a second agency (usually a national) to translate, publish and distribute its materials in another language (Cabanilla and Hargrove, 1986). Co-publication of IRRI materials started in 1973 and it began receiving great interest from many national publishers in the early 80's. In fact, IRRI has been the pioneer in innovating co-publication as a strategy to bridge the language barrier among various groups who usually suffer from serious technological information crisis.

Effectiveness of Co-published Materials

The effectiveness of two materials was evaluated at IRRI's Communication and Publications Department in two separate research projects. The first study by Huque and Hargrove (1987) included two materials: (1) English and Cebuano editions of the book, A Farmer's Primer on Growing Rice by Benite S.Vergard an IRRI scientist in the Plant Physiology Department; and (2) English and Cebuano editions of a poster, National Enemies of Insect Pests of Rice.

The highly illustrated colour poster, measuring 31"×30" was designed originally by the FAO of the United Nations in collaboration of the IRRI to help rice workers and progressive farmers identify beneficial rice field insects. The colours used were: blue, black, yellow and red. The poster was later co-published in 14 languages including some in those of the Philippines, India and Bangladesh.

It has been earlier noted some of the language complexities of the Philippines where the materials were tested. The country, with 7000 islands in the Pacific Ocean and China sea and a population of about 60 millions, has more than 100 dialects. Interestingly half of these have no graphic materials (Jaffery, 1984). About 90 per cent of the population speak eight dialects of which Cebuano ranks top which is closely followed by Tagalog—the dialect of Southern Tagalog province including Metro Manila. English, although a foreign language, is spoken extensively more than any of the local dialects. About 40 per cent Filipinos can speak English (Llamzen, 1969) and about 90 per cent are literate with a national average of grade VI.

The major research questions addressed were: (1) Is the Primer more effective in Cebuano than its original language, English? (2) Is the poster more effective in Cebuano than English? For Primer, effectiveness was operationally defined as knowledge gain, after exposure to it for 45 days, in rice technology. However, for poster, it was knowledge gain in biological control of insect pests of rice, and attitude change toward biological control, both after exposure for the same period.

The research followed the 'one-group pretest-posttest' experimental design. Subjects were 44 rice technicians (extension workers) selected randomly from each of Leyte and Southern Leyte where Cebuano was the major dialect. Two criterion-referenced tests (one for each material) were developed following Bloom's (1956) classical taxonomy of educational objectives to identify the cognition level of the subjects before and after they were given treatment.

The two groups matched in all 39 variables grouped as subject variables ("type-S") situational variables, and treatment variables ("type-E"). Besides, the groups matched in three dependent variables viz., initial knowledge in rice technology, initial knowledge in biological control, and initial attitude toward biological control. All possible measures were taken to control extraneous variables. Leyte technicians were exposed to English edition of the materials in conducting tests and collecting data from the groups were same as those of the materials i.e. English in Leyte, and Cebuano in Southern Leyte.

The results are shown in Figure 1. However, the major findings were as follows:

1. Both groups gained knowledge in rice technology compared to their initial knowledge and the differences were significant at .01 level of probability.
2. Both groups gained knowledge in biological control compared to their initial knowledge and the differences were significant at .01 level of probability.
3. Leyte group gained favourable attitude toward biological control compared to their initial attitude and this difference was significant at .05 level of probability. However, Southern Leyte group gained favourable attitude toward biological control compared to their initial attitude and this difference was significant at .01 level of probability.

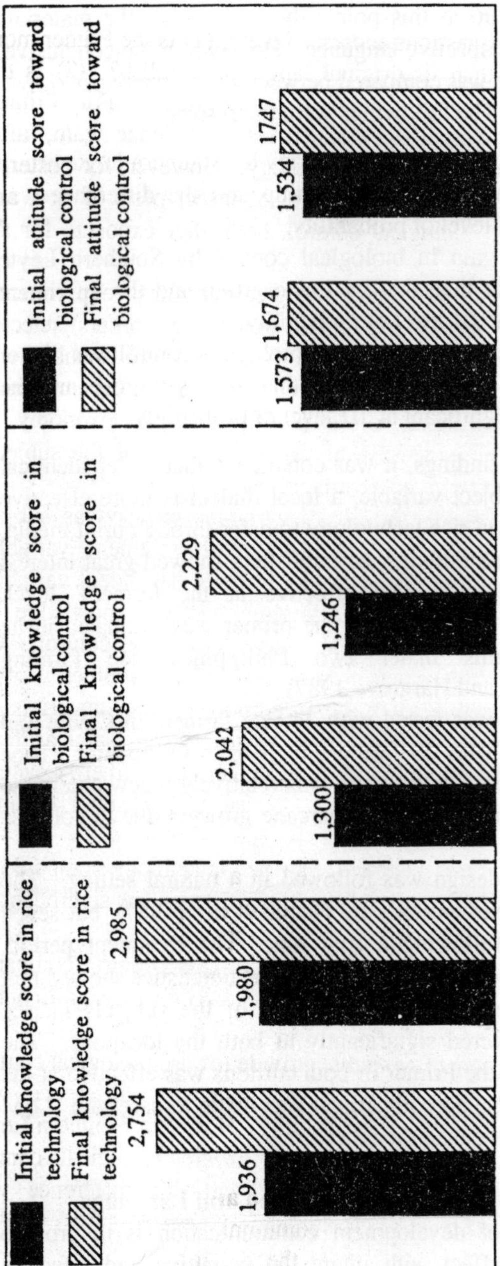

Fig. 1: The effectiveness of A farmer's primer and the Natural enemies poster in English and Cebuano.

The above findings up to this point, show that both the materials were effective in their respective language. However, to be conclusive, the net gain in each case was compared between the two groups—Leyte versus Southern Leyte. The followings were the results:

1. Net knowledge gain in rice technology by Southern Leyte group was higher than that by Leyte group and the difference was significant at .01 level of probability.
2. Net knowledge gain in biological control by Southern Leyte group was higher than that by Leyte group and the difference was significant at .01 level of probability.
3. Finally, net attitude change toward biological control was higher by the Southern Leyte group than that by Leyte group and the difference was significant at .01 level of probability.

Based on the above findings, it was conclusive that under identical field conditions and subject variable, a local dialect is more effective than a foreign language in rice technology transfer through print media. Since the release of these findings, co-publishers showed great interest to learn if the Primer was also effective at the farmers' level. Therefore, in a second study at IRRI, the primer was evaluated in its two co-published forms under two Philippines rice farming communities (Cabanilla and Hargrove 1987).

The Tagalog edition was tested with 34 rice farmers in Cavite and the Hiligaynon edition with 50 farmers in Negros Occidental. Rice farming in the latter speech community was relatively a new enterprise where most farmers were displaced sugarcane growers due to collapse of sugar industry.

The pretest-posttest design was followed in a natural setting. The test was adapted from earlier study by Huque *et al.*, (1987), but seven items were discarded following a trial run. The treatment period, however, was same as before (i.e. 45 days). The t-statistics showed that the means of the pretest and posttest scores of the subjects in rice farming technology differed significantly in both the locations. The conclusion follows that the Primer in both editions was effective at the farmer's level. However, no comparison was made to ascertain if one edition was more effective than the other.

Policy Issues in Development Communication and Language

The basic purpose of development communication is to provide continuous stimuli to affect with intent the cognitive and affective domains of the receivers for favourable responses which must finally elicit stable change in behaviour and action. These need, among others, communication and language policies on a sound footing which most developing countries are lacking. We will take these issues here briefly.

Communication planning oftentimes is not based upon rational thoughts and adequate analysis of clients' need and other relevant issues. The Ethiopian case is a glaring example. The Extension Programme and Implementation Division adopted the 'model farmer approach' which they conceived as having a serious multiplier effect, based on observation of East African experience at a prima facie level. There was no attention given to explore the potential force of mass media, especially the radio on the fallacy that the latter was ineffective possibly because there were only seven receivers per thousand people in the early 70's (Head, 1974).

The frustrating situation appeared when model farmers failed to demonstrate a really multiplier effect in both the countries. In East Africa model farmers progressed well in adopting technology in their own farms, but they could not show diffusion impact on their fellow farmers. On the other hand, Ethiopian model farmers were not even prepared to work hampering their own farming business and secondly, there was no reward for their being the 'models' (Schulz, 1981). Almost the same policy of Contact Farmers (CF) approach is deeply rooted into the rural extension communication network in the developing countries where T&V system had been adopted since the 70's. Through a regular schedule of training for agents and their follow-up visits to the CFs, both groups seem to have been abreast of the latest farm technologies. The philosophy behind concentrating effort on the CFs is that each of them must communicate the 'impact points' to at least 10 other farmers in the community. But this 'two-step' flow of technology diffusion seems to be inoperative in most cases.

In Bangladesh, for example, in the first step a regular visit by Block Supervisors is accounted to only 20 per cent of their CFs, occasional (3-5 times in a rice season of about four months) to 55 per cent and rare or never to rest 25 per cent. The communication skills of the BSs along with other extension personnel in the organizational hierarchy were also found unsatisfactory (Karim, 1987). Keneyan extension workers of T & V system also lack the needed communication skills (Gatere, 1988). However, whatever the CFs gain from the BSs, sharing of technologies with fellow farmers is minimal in the second step (Karim, 1987). Surprisingly, the CFs are reluctant to improve the socio-economic status of the latter group and a significant difference between the groups exist in agricultural knowledge, yield of rice, and adoption of farm technologies (Halim, Ali and Islam, 1986). A similar gap in T & V message dissemination between CFs and follower farmers are observed in Thailand where 66 per cent of the CFs do not perform any role as link men between extension agency and the vast clientele group (Chumsri, 1982). In Nepal, messages of the T & V are found to be progressively lost at each point along a top-down hierarchical set-up i.e., Subject Matter Specialists—extension workers—contact farmers—follower farmers (Baral, 1981).

A similar information diffusion gap was observed long before the T & V system was introduced. In the Philippines during early stage of the Green Revolution with miracle rice, several studies showed evidence of 'information hoarding' among the first recipients of new rice technology. They wanted to keep the new technology as 'farm secret' and that many of them even did not discuss the technology with their wives simply because the woman has no secret (Castillo, 1975).

Attempts to strengthen print media use on a policy prescription level have often been ignored on the assumption that most farmers in developing countries are illiterate. This seems an invalid preposition as the Bangladesh case tends to suggest. The national statistics show that only 25 per cent of the 'population' are literate. A random probing into the masteral theses completed on diffusion of farm technologies at BAU shows surprisingly that in most studies majority (over 50 per cent) of the farmers are literate with various levels, however, primary level mostly dominate over any other levels. The fact that literacy rate of farmers far exceed the national statistics could be attributed to these factors: (a) active farmers belong to a particular age group, let us say 20-50 years, (b) more of the younger farmers are somewhat literate than they were in the past, and most importantly, (c) socially higher value on education is given to males than females and thus males dominate by more than two-folds than females in the literacy rate. Subject to further validation, it can safely be asserted that where the great majority of the farmers are males, most of them have some kind of education in those countries where the national literacy figure exceed to 25 per cent. The vital issue from communication view point, however, is at what literacy level a typical extension publication can they read with satisfactory comprehension of the technological messages? Answer to this can only be had by readability research of the extension publications which seriously lack in developing countries. On a tentative basis, it may be observed that print media if well designed based on specific communication purpose, farmers' need and readability level, can do a better job than these do today in the transfer of technology.

Lack of adequate general and technical vocabularies is common to most languages in the world. Further, very few of the world's language possess a satisfactory orthographics. Deliberate planning therefore should get prime consideration. Previous review on the effectiveness of co-published materials does not really solve the linguistic barriers where the latter bring about noises of many kinds. It simply suggests as a policy matter that communicating the clients in their known language would be more helpful in disseminating information than doing it in any other language.

Literature, indicate that language planning has been attempted in many occasions both on regional and national levels with limited success when judged against benefited from the common mass. For example, effort to improve communication by producing a standardized lingua franca from several related languages has not been successful in the highly multi-lingual communities of Africa (Doob, 1961). The solution to the acute dearth of technical vocabulary in scientific areas is not seen an easy task. In 1961, for example, the Institute of language (INL) of the Philippines came up with a list of vocabularies in ten specialized fields e.g., biology, law, traffic etc. The lone volume consisted of only 171 pages where only 70 biological terms were developed. Tagalog equivalents were given for pollen, stamen, and pollination as *polen, istamen,* and *paglipat ng polen,* respectively. But vocabularies were still to be suggested for genes, chromosomes, diploidy and triploidy (Roces, 1968). Under such a vocabulary constraint, it was not possible by the INL to translate in Tagalog an English version of the booklet, "How to Grow Seedless Watermelons", sent by the Bureau of Plant Industry—the Extension Department of that country.

Indian linguists, especially in Hindi, however, showed a remarkable progress in their language development task. The 1951 census reported a total of 845 languages and dialects in the country and Hindi, spoken by 42 per cent of the population, became the principal official language (Llamzon, 1973). The Hindi literacy scholars in collaboration with the members of the *Nagarik Procharani Sabha,* prepared six volumes of encyclopedia in 10 years since they were assigned the task in 1956.

The Board of Scientific Terminology, constituted in 1950, came up with 290,000 new terms in 1963 out of a planned target of 350,000.

The policy followed by the literacy scholars was to retain Hindi's purity from contamination which finally diverged the language from the tongues of the common men. Thus, 'smoking is prohibited' took the classic form as 'dhumrapaan varjit hai' rather than the popular form 'sigret piinaa manaa hai' (Gupta and Gumperz, 1968). The same policy in Bengali language development is being followed more seriously since Bangladesh got her independence in 1971. The works of the scholars seem to suggest that these might eventually make Bengali a language of the elite class. Bengali development of West Bengal state in India seems also following the same trend. The vital question here—how language development fulfils the objective of transmitting technologies to the vast majority of the clientele group? Is the work of classical language development supportive to technology transfer process? Or, it creates noise in some other forms?

In the absence of proper language policy, the use of language per se is oftentimes irrational compared to the demands made upon it by the people and the community that know language. Nigeria, for example, has 178 languages. Broadcasting seems to use local languages more than the print, ironically however, English dominate broadcasting than the local languages, especially in Southern Nigeria in all programme items, e.g., news, sports and education. Further, of the three major Nigerian languages—Hausa, Yoruba, and Igbo, only the first two are most utilised in the broadcast media (UNESCO, 1980). Similarly, broadcast for rural people in the major local languages is insignificant in Ethiopia (Schulz, 1981).

Recommendations

Based on our analyses and understanding, we would first define the social context on which we wish to recommend some action plans to strengthen development communication in the Third World countries.

First, development communication should be conceived as one of the components of agricultural and other changes in that it is necessary but not sufficient alone to affect the processes. Second, the major barrier to development today in most Third World countries is not the generation of technology, but its diffusion (Burki and Goering, 1977). Third, surprisingly enough, developing countries spend three times higher in extension than in research, while developed countries do the opposite—invest more in research than in extension (Howell, 1982). Fourth, our analyses show that there exists information crises which gradually aggravate among the one-to-one level of the research—linking-client systems. Fifth, within the social framework of particular client system, access to technological information has been a monopoly of the "minority fortunates" against the vast "majority underprivileged."

Within the purview of the above context and based on our detail analyses, the following recommendations could be made.

1. Effective communication strategies, backed by empirical evidence from developing countries, should be planned and implemented in the transfer of technology in agriculture and other social change programmes. Our concept of an *effective communication strategy* rests upon the following bases:

(a) It is deliberately a well planned activity.
(b) It is development oriented—some thing it purports to change clients' socio-economic status and level of living.
(c) It is objectively specific—values are laid on what it wishes to accomplish.
(d) It is agro-ecologically and socio-linguistically situation specific.
(e) Its cost is reasonably affordable by the organization concerned.
(f) It is tailored upon manipulating the variables of the basic elements of communication in the most possible favourable direction.
(g) Finally, it is reasonably effective to face competition from other communication strategies that are being launched by any other organization with similar goals, plans and target clientele.

2. Adequate information dissemination cannot be ensured to all social strata of the client system by largely depending upon the interpersonal media. Experience shows that the 'hypodermic needle model' does not operate only in the advanced countries. The complete boycott of food items imported from an European country (let us say it is country E) by the Bangladeshis in early 1987 is a strong case of evidence.

Following the Charnobyl nuclear plant disaster of the USSR in April 1986, messages continued to disseminate for about a year only through the mass media (radio, television, newspapers and weekly magazines) that E's dairy products imported to Bangladesh had been contaminated with radioactivity to a level hazardous to human health. The E country also launched a campaign refuting the alleged radioactivity level also through mass media. This led to a heated "communication war" between government sources and E's embassy office in Dhaka.

Public boycotted the milk products of all European countries including E. The sweetmeat business came to a complete halt and lemon tea became very popular. Also milk products were replaced by sweet potato and carrot as ingredients for sweetmeat preparation and the latter also got some acceptance. However, the controversy gradually subsided with no clear follow-up action from the government source, and the people, rightly or wrongly, started to consume these foods again as before after about six months of discontinuance.

Communication specialists would argue that government messages contained 'threat appeal' so that public responded to it readily. This may be partially true on peoples' rejection behaviour, but the power of mass media was clearly evident in a country where direct access of the public to mass media has been very limited. Thus, mass media can perform a better role in technology diffusion than what these do today. Therefore, planned efforts to introduce more of the mass media strategies, proven effective by experiments, are highly recommended.

3. Radio and television programmes in developing countries have been greatly urban oriented with entertainmental bias and less educational for the rural masses. Therefore, more educational programmes through these two potential media should be launched with different strategies for the rural clientele.

4. The volume of print media published by the extension agencies seems to be inadequate compared to client's demand. Also there seems no distribution policy for these to the farmers. Therefore, adequate number of print media should be published followed, by a workable distribution mechanism so that these really reach the deserving clientele.

5. More thoughtful agricultural publications (e.g., books, magazines, guides etc.) should separately be prepared by the extension agencies both for change agents and their clients. A good marketing mechanism needs to be developed through which the materials would reach to the consumers at a subsidized rate.

6. IRRI's experience on copublication, as a strategy of breaking the language barrier, seems to be successful. To be conclusive, more copublication researches with cross-cultural subjects and languages need to be undertaken. However, it may be recommended that major publications be copublished in local dialects rather than doing it in the classical form of the language. Depending on the characteristics of a speech community and its population, languages of the common men be used in development messages through local radio, television and print media.

7. The dominant role of the BSs and CFs in the interpersonal communication network under the T & V system would continue in an unpredictable future. Therefore, their training schedules must emphasize developing communication skills along with other technical subject matters.

8. Research in communication seems to be inadequate in most developing countries. Further, some research areas remain untouched. Research may therefore be undertaken with special emphasis in these directions:

(a) Change agents' accessibility to various communication technologies in conducting field extension works.
(b) Proficiency level of change agents and clients in the major language in which they communicate.
(c) Functional literacy of the farmers and its level in which an average extension print media can satisfactorily be decoded by them and comprehended.
(d) Readership behaviour of and attitude toward professional publications by change agents and clients.
(e) Evaluation of print media by readership research following Flesch formula and/or close technique.
(f) Content analyses of development messages published in newspapers and farm magazines.

9. Development communication as a distinct field of academic pursuit is yet to grow in many Asian and African countries. Also there is a serious lack of trained manpower in the field. Communication planning, therefore, suffers from serious set-backs and oftenly it is done on an ad-hoc and trial-run basis. Therefore, more exposures in development communication should be integrated into the course curricula of agricultural degree programmes. Post graduate studies in development communication should also be introduced wherever possible. Extension organizations should not only be manned by technical subject matter specialists, but also by communication specialists, however, with basic background in agriculture at their bachelors' study programme.

10. Each country must have its own development communication and language policy linked closely with its general communication policy. The technical communication planning should rather be conceived as a subsystem of the general communication system. The language policy for technical communication should be based on the following issues:

(a) Collection of key terms in major languages, dialects and screeming out of the misleading ones which create noise in encoding development messages.
(b) Reduction of excessive synonymous terms into a sizeable limit to check noise due to possible entropy for the encoder.
(c) Development of technical terms either in the same language or borrowed from most related language.
(d) Development of new vocabulary where the same term refers to antonym.
(e) Developing graphic materials where the language calls for.
(f) Development of easy and simple orthography wherever feasible.
(g) Development of less number of homonyms to avoid noise that may occur in oral communication.

11. Putting the above recommendations into action plan is a gigantic and complex task which most developing countries would be unable to do. For, these need communication resources of many kinds which are lacking in most of them.

It is thus strongly recommended that collaborative research and action plan be undertaken by several countries with common agro-climatic and socio-linguistic characteristics. For example, collaborative works on language issues may be initiated under SAARC sponsorship between India and Bangladesh for Bengali language; and India and Pakistan for Punjabi, Hindi-Urdu and Kashmiri languages and so on.

The major steps that could be followed in solving problems of technical language are: (a) collection of key terms in major crops, weeds, fertilizers, insect pests through survey research design over widely dispersed locations, (b) identification of the noise-related issues inherent in that language, (c) solution of the linguistic problems by a team of experts consisting of subject matter specialists (e.g. agronomists, entomologists, plant pathologists etc.), extension and communication specialists and learned linguists—all need to be selected from the countries concerned.

The "Asian Network on Development Communication by Agricultural Universities" has been established in May 1988 under the sponsorship of the FAO's Regional Office for Asia and the Pacific (RAPA). Most SAARC and ASEAN countries have been included in the network. The network is proposing to undertake intercountry research projects in development support communication. Some of the problems outlined in this article may be taken in such collaborative research endeavour.

Finally, we would recommend that a development communication division be established in each of the 13 CGIAR research institutes. We believe that technologies alone are not enough without their planned diffusion in the search of a brighter world for tomorrow.

Acknowledgement

The author expresses his deep gratitude to Professor Md. Anisur Rahman, Department of Management, University of Rajashahi, for his inspiration and assistance by providing some valuable literature for writing the paper.

References

AGRIS. 1973 International Information System for the Agricultural Sciences and Technology. F.A.O.

Ahad, A., M. Roy, and M.A. Sardar. 1987. Krishi Keetbijnan (Bengali). Mymensingh: Ali Press.

Ahmed, M. 1987. A New Information Order. Bangladesh Quarterly. 7(1)

Allen, M.B. 1982. Listening: The Forgotten Skill. N.Y.: John Willey & Sons Inc.

Awa, N.E. and L.V. Crowder, Jr. 1978. How extension Stacks Up. Journal of Extension. Vol. 17 (March-April).

Ball, N. 1981. World Hunger: A Guide to the Economic and Political Dimensions. California : ABC -CLO, Inc.

Baral, J.R. 1981. The Dynamics of Field Implementation of the Training and Visit System in Parsa District of Nepal. Unpub. Ph.D. thesis, UPLB.

Barry, G.J. Bronowski, J. Fisher, and J. Huxley. 1965. Communication and Language: Networks of Thought and Action. N.Y.: Doubleday & Co. Inc.

Bernardo, F.A. 1981. Catering to the Information Needs of Extension Workers. In Agricultural Information to Hasten Development. Agril. Inf. Bank for Asia.

Bhuiyan, M.S.I. 1988. Use of Communication Media by the Farmers in the Adoption of Selected Improved Farm Practices in Rice Cultivation. Unpub. M.Sc. (Ag. Extn. Ed) Thesis, BAU.

Bloom, B.S. (ed.). 1956. Taxonomy of Educational Objectives: Handbook I, Cognitive Domain, N.Y.: David McKay Co., Inc.

Burki, S.J. and T.J. Goering. 1977. A Perspective on the Foodgrain Situation in the Poorest Countries. World Bank Staff Working Paper No. 251.

Busch, L. (ed.) 1981. Science and Agricultural Development. N.J.: Allanheld Osmun Pub.

Cabanilla V.L., and T.R. Hargrove. 1986. Copublication in the Third World: Breaking the Language Barrier. Scholarly Publishing. University of Toronto Press.

Caslillo., G.T. 1975. All in a Grain of Rice, Los Banos: SEARCA.

Chumsri, P. 1982. Operationalising the Training and Visit System: The Case of the National Agricultural Extension Project in Thailand. Unpub. Ph.D. thesis, UPLB.

Deutsch, K.W. 1966. Nationalism and Social Communication: An Inquiry into the Foundations of Nationality. London: The MIT Press.

Deutschman, P.J. *et al.*, 1968. Communication and Social Change in Latin America: Introducing New Technology. N.Y.: Frederick A. Praeger, Pub.

Dey, P.K. 1968. Relative Effectiveness of Radio and Television as Mass Communication Media in Dissemination of Agricultural Information. Unpub. M.Sc. thesis, IARI.

Doob, L.W. 1961. Communication in Africa: A Search for Boundaries. London: Yale University Press.

Encyclopedia Britannica. 1971. Chicago: Encyclopedia Britannica Inc.

Escarpit, R. 1966. The Book Revolution. London: George C. Harrop & Co. Ltd. and UNESCO.

Fishman, J.A. 1968. Some Contrasts Between Linguistically Homogeneous and Linguistically Heterogeneous Politics. In Fishman J.A., C.A. Ferguson, and J.D. Gupta (eds). Language Problems of Developing Nationals. N.Y.: John Willey & Sons.

Gatere, K. 1988. Higher Agricultural Training in Keneya: Fitting the Task to the Needs. In Training for Agriculture and Rural Development. Rome: FAO

Glatthorn, A.A., C.W. Kreidler, and E.J. Heiman. 1971. The Dynamics of Language. London: D.C. Heath and Co.

Gupta, J.D. and J.J. Gumperz. 1968. Language, Communication and Control in North India. In Fishman. J.A., C.A. Ferguson and J.D. Gupta (eds). Language Problems of Developing Nations. N.Y.: John Willey & Sons.

Halim, A., M.M. Ali and M.M. Islam, 1986. The Effect of T & V Extension System on Contact and Non-contact Farmers. Bangladesh Journal of Extension Education 1(1).

Hancock, A (ed.). 1984. Technology Transfer and Communication. Paris: UNESCO.

Havelock, R.G. 1975. Planning for Innovation (5th printing). Institute for Social Research, the University of Michigan.

Head, S.W. (ed.) 1974. Broadcasting in Africa: A Community Survey of Radio and Television. Philadelphia Temple University Press.

Hitti, P.K. 1961. History of the Arabs. London: Macmillan & Co. Ltd.

Howell, J. 1982. Managing Agricultural Extension: The T and V System in Practice. London: Agril. Admn. Unit, Overseas Dev. Institute.

Huque, M.M. 1972. Information Sources Used by the IRRI Rice Growers of Thakurgoan Thana Cooperative Tubewell Project. Unpub. M.Sc. (Ext. Ed.) thesis, BAU.

Huque, M.M. 1982. Masagana 99 Farmers' Perception of the Effectiveness of Communication Media and their Use in Bay, Laguna. Unpub. M.S. thesis, UPLB.

Huque, M.M. 1986. The Relative Effectiveness of two Extension Publications in English and Cebuano on Change Agents' Cognitive and Affective Domains in Rice Technology Diffusion. Unpub. Ph.D. Thesis, GAUF.

Huque, M.M., T.R. Hargrove and P.L. Esteban. 1987. The Effectiveness of Language: A Study of Two Rice Extension Publications in English and Cebuano Among English-Speaking Change Agents in the Philippines. IRRI Research Paper Series No. 126.

Jamias, J.F. 1979. Towards Effective Rural Communication: Some Insight From Research. Mimeographed paper, Dept. of Dev. Com., UPLB.

Jernudd, B. 1968. Linguistic Integration and National Development: A Case Study of the Jebel Marra Area, Sudan. In Fishman, J.A., C.A. Ferguson, and J.D. Gupta (eds). Language Problems of Developing Nations. N.Y.: John Willey & Sons.

Juaquin, T.P. 1978. The Translatability from English to Philippine Terms Used in Rice Production: An Exploratory Study. Unpub. M.S. thesis, UPLB.

Juliano, V.A. 1981. The Effectiveness of a Vernacular Farmer's Bulletin in Communicating Agricultural Information. Unpub. M.S. thesis, UPLB.

Karim, A.M.A. 1987. Communication Strategies and Extension Teaching Methods. Bangladesh Journal of Extension Education 2 (Special Seminar Issue.)

Knox, J.W.. 1962. Relative Value of Mass Media. Abstract in Review of Extension Research, USDA Ext. Ser. Cir. 541.

Latif, M.A. 1974. How Farmers Receive Agricultural Information. Unpub. M.Sc. (Ag. Ext. Ed.). thesis, BAU.

Lele, U.J. 1975. The Design of Rural Development: Lessons from Africa. London: The John Hopkins University Press.

Liuton, R. 1936. The Study of Man: An Introduction. New York: D. Appleton Century Co., Inc.

Lionberger, H.F. and P.H. Gwin. 1982. Communication Strategies: A Guide for Agricultural Change Agents. Illinois: The Interstate Printers & Publishers, Inc.

Litsinger, J.A., E.C. Price, and R.T. Herrara. 1980. Small Farmer Pest Control Practices for Rainfed Rice, Corn and Grain Legumes in three Philippine Provinces. The Philippine Entomologist. 4(1-2).

Llamzon, T.A. 1969. Standard Filipine English: Language Research Series. Ateneo de Manila University.

Llamzon, T.A. 1973. The Integrative Function of Language: Do We Need a National Language to Unite Us? Philippine Studies. 25 (Third Quarter).

Mattar, A.C. 1975. Arab Linguistic Diversity and Mass Communication. In Getting the Message Across. The Unesco Press.

Mitchell, T.R. 1982. People in Organisation: An Introduction to Organizational Behaviour (2nd ed.).

Morris, J.M. and E.A. Elkins. 1978. Library Searching: Resources and Strategies. N.Y.: Jeffrey Norton Pub.

Nekby, B. 1971. CADU: An Ethiopian Experiment in Developing Peasant Farming. Stockholm: Prisma Pub.

Pascasio, E.M. 1964. Language: An Aid to Cross-Cultural Understanding. Philippine Sociological Review. 12(1-2).

Potter, J.N., D.H. Barden, and E. McMaugh. 1974. Extension and other Duties. Journal of the Australian Institute of Agricultural Science. 3(40).

Rahman, M.M. 1974. An Investigation into the Use of Communication Sources by the Registered Jute Seed Growers of Meherpur Thana. Unpub. M.Sc. (Ext. Ed.) thesis, BAU.

Roces, A. 1968. The Choice of Languages of Instruction. Philippine Journal for Language Teaching. 6(1-2).

Rogers, E.M., J.R. Brown and M.A. Vermilion. 1977. Radio Forums: A Strategy for Rural Development. In World Bank Staff Working Paper No. 266.

Rogers, E.M. 1980. Communication Technology: The New Media in Society. N.Y.: The Free Press.

Sandhu, A.S. 1970. Characteristics, Listening Behaviour and Programme Preferences of Radio Owning Farmers in Punjab. Unpub. Ph.D. thesis, New Delhi: IARI.

Schulz, M. 1981. Extension Service in Ethiopia: Strategies before and after Revolution. In Crouch, B.R. and C. Shamkariah (eds). Extension Education and Rural Development. Vol. 2 John Willey & Sons Ltd.

Singh, J. 1971. A Study of the Factors Influencing the Viewing Behaviour of the Farmers Towards Agricultural Programme on Television in Delhi Villages. Unpub. Ph.D. thesis, New Delhi: IARI.

Singh, K.N. 1981. The Need for a Communication Strategy for Rural Development. In Crouch, B.R. and S. Chamala (Eds) Extension Education and Rural Development. Vol. 2. John Willey & Sons Ltd.

Smith, Jr. D.C. 1983. Keynote Address. In Copublication: Strategies for Multilanguage Publication in Agriculture. Los Banos: IRRI.

Spaddley, J.P. 1979. The Ethnographic Interview. N.Y.: Holt, Rinchart and Winston, Inc.

UNESCO. 1957. Scientific and Technical Translating and Other Aspects of the language Problems Paris: Documentation and Terminology of Science.

UNESCO. 1965. Books for the Developing Countries: Asia, Africa, Paris.

UNESCO. 1980. Communication Policies in Nigeria. Paris.

UNESCO. 1981. Farm Broadcasting in Selected Countries of Asia and Pacific. Bangkok.

Viloria, R.V. 1975. Bibliographic Control of Agricultural Information and Language Problems in Southeast Asia. In Proceedings of a Workshop/ Seminar on Regional Cooperation in Agricultural Information. Los Banos: SEARCA.

Watcon, J. 1985. English, the International Language of Science. CBE Views 8(2).

3

A Transfer Technology Programme: Whether A Myth ?

D. Das Gupta and Subir Bhattacharya

It was held that the body of knowledge generated at the research stations usually followed a slow rate in the process of their transfers to the actual clients. Transfer technology programmes are specifically directed towards carrying the duty to communicate these knowledge to the users. A lot of transfer technology models have been developed to expose the standards that can be followed for successful technology transfer. These models were adopted/implemented either partially or in some combinations since India's independence. Success of these structures actually rest on their practical applicability. Hence, a thorough analysis of the impact of transfer technology programmes on effective amelioration of socio-economic front of the clientele and using the resultant feed-back to modify these models in pragmatic sense are obviously imperative. Although we have a variety of transfer technology programmes launched at different points of time since independence, but time and space permitted us to restrict our investigation on the basis of a specific one. The present investigation was carried out with reference to the Lab to Land Programme (LLP) : a transfer technology programme launched by ICAR in 1979 with a view to diffuse the latest proven economically viable technology among the poorest of the poor. The specific objectives of this study were as follows :

(i) to assess the impact of LLP on productivity; and
(ii) to examine the distributive effects of LLP on small farm economy.

Now, the systematic presentation was done in three sections. In the first section a theoretical model and the statistical methods followed in this study, were described. In the second section the test results were discussed. Finaliy, sonie conclusions were drawn in the third section.

I

An Econometric Model to assess the Impact of Technology Transferred on Gross Cropped Area, Cropping Intensity and Productivity :

The model starts by specifying the following notations:

P_j = productivity for the jth farmer;
Y_j = total volume of agricultural output of the jth ; farmer
N_j = net cropped area for the jth farmer;
A_j = gross cropped area for the jth farmer;
C_j = cropping intensity of the jth farmer;
m_j = family size of the jth farmer;
$Q_j = Y_j/m_j$ = per capita household output;
t_j = technology for the jth farmer;

$T_j = \frac{t_{aj} - t_{bj}}{t_{bj}}$ = technology transferred to the jth farmer;

$X_j = \frac{C_{aj} - C_{bj}}{C_{bj}}$ = rate of change of cropping intensity for the jth farmer;

$R_j = \frac{A_{aj} - A_{bj}}{A_{bj}}$ = rate of change of gross cropped area for the jth farmer;

b = before 1982 ; and
a = after 1984.

Let us assume that there exists a functional relationship among productivity, technology transferred, rate of change of cropping intensity, rate of change of gross cropped area and per capita household output. On specification, it is further assumed that productivity is an increasing function of the other variables. Thus, one can write

$$P = F(T, R, X, Q) \qquad (1)$$

where,

$$\frac{dP}{dT} > 0, \frac{dP}{dR} > 0, \frac{dP}{dX} > 0 \text{ and } \frac{dP}{dQ} > 0$$

Now, it is plausible to break (1) into a simultaneous equation model (to make the relationship more detailed and specific) in the following manner :

$$R = \phi\,(P) \quad (2)$$
$$X = \psi\,(P, T) \quad (3)$$
$$P = \theta\,(R, X, Q) \quad (4)$$

For simplicity let us lay down the linearity assumption, and with the aid of it the statistical form of the model (2) - (4) becomes

$$R = a_o + a_1 P + U_1 \quad (5)$$
$$X = b_o + b_1 P + b_2 T + U_2 \quad (6)$$
$$P = c_o + c_1 R + c_2 X + c_3 Q + U_3 \quad (7)$$

where,

$$U_1 \sim N\,(0, \sigma^2_{u1})$$
$$U_2 \sim N\,(0, \sigma^2_{u2})$$
$$U_3 \sim N\,(0, \sigma^2_{u3})$$

and U_1 s have other properties of the residual error term. The model (5) - (7) contains three endogenous variables: R, X and R. The reduced-form model for (5) - (7) are then worked out as

$$R = \pi_{10} + \pi_{11} Q + \pi_{12} T + V_1 \quad (8)$$
$$X = \pi_{20} + \pi_{21} Q + \pi_{22} T + V_2 \quad (9)$$
$$P = \pi_{30} + \pi_{31} Q + \pi_{32} T + V_3 \quad (10)$$

where,

$$\pi_{10} = a_0 + \frac{a_1 \left(c_0 + c_1 a_0 + c_2 b_0\right)}{1 - a_1 c_1 - c_2 b_1}$$

$$\pi_{20} = b_0 + \frac{b_1 \left(c_0 + a_1 c_1 + b_0 c_2\right)}{1 - a_1 c_1 - c_2 b_1}$$

$$\pi_{30} = \frac{c_0 + c_1 a_0 + c_2 b_0}{1 - a_1 c_1 - c_2 b_1}$$

$$\pi_{11} = \frac{a_1\, c_3}{1 - a_1 c_1 - c_2 b_1}$$

$$\pi_{21} = \frac{b_1 c_3}{1 - a_1 c_1 - c_2 b_1}$$

$$\pi_{31} = \frac{c_3}{1 - a_1 c_1 - c_2 b_1}$$

$$\pi_{12} = \frac{a_1 c_2 b_2}{1 - a_1 c_1 - c_2 b_1}$$

$$\pi_{22} = b_1 + \frac{b_1 c_2 b_2}{1 - a_1 c_1 - c_2 b_1}$$

$$\pi_{32} = \frac{c_2 b_2}{1 - a_1 c_1 - c_2 b_1}$$

$$V_1 = u_1 \frac{a_1 (c_1 u_1 + c_2 u_2 + u_3)}{1 - a_1 c_1 - c_2 b_1}$$

$$V_2 = u_2 \frac{b_1 (c_1 u_1 + c_2 u_3 + u_3)}{1 - a_1 c_1 - c_2 b_1}$$

$$V_3 = \frac{c_1 u_1 + c_2 u_3 + u_3}{1 - a_1 c_1 - c_2 b_1}$$

V_i s are the stochastic error terms having usual properties of the residual error term.

The reduced - form parameters measures the total effect, direct and indirect, of a change in the predetermined variables on the endogenous variables, after taking account of the interdependences among jointly dependent endogenous variables. The impact of technology transferred on productivity, rate of changes of cropping intensity and gross cropped area can now be observed from π_{32}, π_{22} and π_{12} respectively. If these are found to be significant in the fitted model of the type (8) - (10) then one will be able to generalize about the impact of technology transferred on the endogenous variables.

As regards the impact of LLP on income distribution, the New Co-ordinate System Approach (NCSA) [1] was followed. According to this approach Lorenz curve is defined as

$$\eta_1 = a\, \pi_t^{\alpha} (\sqrt{2} - \pi_t)^{\beta} \; ; \; a, \alpha, \beta > 0 \qquad (11)$$

where, η is the length of the ordinate from, say P (a point on the Lorenz curve) on the Egalitarian line and π is the distance of the ordinate from origin along the Egalitarian line. (11) can also be written as

$$\log y_t = a' + \alpha \log r_t + \beta \log\left(\sqrt{2} - r_t\right) + W_{1t} \tag{12}$$

where, r_t and y_t are efficient estimaters of π_t and η_t ; $a' = \log a$; and W_{1t} is the residual error term.

On the basis of (12) Gimi Concentration Ratio (GCR) is obtained by the following formula :

$$GCR = 2\int_0^{\sqrt{2}} a\, r_t\left(\sqrt{2} - r_t\right) dr_t \tag{13}$$

which on simplified form can be written as

$$= \text{anti} - \log\left[1{\cdot}4142 \log \hat{a} - 0{\cdot}9241\left(\hat{\alpha} - \hat{\beta}\right)\right] \tag{14}$$

where, $\hat{a}, \hat{\alpha}$ and $\hat{\beta}$ are the estimated values of the parameters in equation (12).

Two other measures of inequality were also computed. As per NCSA their formulae can be written as given under :

The Relative Mean Deviation = R

$$= \left(\sqrt{2}\right)^{1+\alpha+\beta} \frac{a\alpha^{\alpha}\beta^{\beta}}{(\alpha+\beta)^{\alpha+\beta}} \tag{15}$$

Eltito and Frigyes's [2], $v^1 = 1 - \dfrac{\left(\sqrt{2}-\pi^*-\eta^*\right)}{\left(\sqrt{2}-\pi^*+\eta^*\right)} \dfrac{\left(\pi^*-\eta^*\right)}{\left(\pi^*+\eta^*\right)}$ (16)

where, $\pi^* = \dfrac{\sqrt{2}\,\alpha}{\alpha+\beta}$ and $\eta^* = \left(\sqrt{2}\right)^{\alpha+\beta} \dfrac{a\alpha^{\alpha}\beta^{\beta}}{(\alpha+\beta)^{\alpha+\beta}}$

II

The prime data base of this study consisted of a random sample of 105 target group farm families of LLP of West Bengal who were associated with the crop production enterprize during its second phase (1982 - '84).

In order to avoid any conceptual variation in the exact meanings of some of the crucial variables used in this study, it was considered necessary to clarify the same.

Technology: A great deal of confusion surrounding the discussion of technological change is no doubt due to the ambiguity that surrounds the term in its conventional usage. It is frequently used in a qualitative sense from 'less' to 'more' while such cardinally quantifiable usage of the term is sufficient for many purposes, it presents many problems in many other cases. Obviously, the usage of the term 'technology' involves similar problems. Technology, as considered in this study, is simply improved agricultural practices. It is measured by the total score obtained by a farmer from different practices under the assumption that the adoption of each and every practice results in a score '1' for him or '0' for non adoption. This is, in fact, a quantification.

Income: In the existing literature, several definitions of income are available. But most of these concepts can not be easily usable in an empirical study because most of the entities involved do not permit objective quantification. Therefore, a simple concept have been adopted for this study. It consisted of the sum of values of different agricultural outputs taken together plus income from secondary sources.

The model (8) - (10) was then fitted to the data and placed in Table 1. The R^2 values in case of all the equations were found to be quite higher. This indicated that the model gave an overall good fit to the data. The parameter values of 'technology transferred' were found to be significantly positive for all the three dependent variables. Thus, it could be said that 'technology transferred' had positive 'multiplier effect' on productivity, gross cropped area and cropping intensity. Major implications of this effect were that available land was used more intensively and economically which resulted in enhancing the overall productivity.

Table 1: The Fitted Reduced-Form Model

Dependent Variables	*Inter-cept*	*Techno-logy transferred*	*Per capita household output*	R^2
Rate of change of gross cropped area	1.1417	0.6295 (2.141)*	0.8896 (1.115)	0.82582
Rate of change of cropping intensity	1.7944	0.9256 (2.743)**	0.9713 (1.702)	0.78981
Productivity	17.4277	0.8711 (2.562)*	1.6546 (1.994)*	0.77394

Figures in the parentheses indicate 't' values

** Significant at 1% level, * Significant at 5% level.

This could be attributed to the exposure of the programme and hence, it indicated an encouraging impact of LLP on the productivity front. This is one part of the results. The other part of it was reflected in Table 2. Lorenz curves were fitted for both the time periods: before 1982 and after 1984. The fitted curves for both the periods had some common characteristics: (i) both fell below the egalitarian line (since $\hat{a} > 0$) and (ii) both were showed towards (0, 0) [Since, $\hat{\alpha} > \hat{\beta} \Rightarrow \hat{\alpha}/\hat{\beta} - 1) > 0$]. The results amply revealed that the income distribution for small and marginal farmers were skewed towards the lower income classes - this skewness was found to be valid for both the points of time. The more important result obtained here, was that the magnitude of the parameters (i.e. $\hat{a}$, $\hat{\alpha}$ & $\hat{\beta}$) were higher for period 'after 1984'. The Lorenz curve for this period was observed to be away from the egalitarian line in comparison to that for the period 'before 1982'. This indicated an increase in the inequality of income among the target group. This was further supported by the increase in GCR, v' and R from 0.0983077, 0.50992 and 0.23042 respectively (in before 1982) to 0.1721394, 0.70654 and 0.27591 respectively (in after 1984). Thus, it could easily be said that LLP had resulted in worsening the income distribution of the target group. Technology involved scale-bias even within a group of poor peasants. Similar results were obtained by Kannathal [3] and Sarma [4]. Although LLP was specifically meant for the poorest of the poor, it failed to fulfil the welfare objective by way of widening the disparity in income among the target group.

Table 2: **Fitted Lorenz Curves and Inequality Measures**

	Lorenz Curve			*Inequality Measures*			
Coefficient estimates	$\hat{a}$	$\hat{\alpha}$	$\hat{\beta}$	R^2	*GCR*	V^1	*R*
Before 1982	0.2049	0.4807 (4.56)**	0.4164 (3.14)**	0.876	0.09831	0.50992	0.23042
After	0.3902	1.0406 (6.66)**	0.8392 (12.77)**	0.985	0.17213	0.70654	0.27591

Figures in the parentheses indicate 't' values
** Significant at 1% level.

III

In the concluding section a Paradigm (Fig. 1) had been developed to expose various aspects of this investigation. A transfer technology programme line LLP were meant for translating the technologies generated through research at a certain point of time with a definite socio-cultural background for the benefit of the clients. Through effective implementation of a transfer technology programme, adoption/ retention of advocated technologies may occur in the social system with definite changes in different dimensions. The various dimensions of positive changes conceived in this study were changes in productivity, gross cropped area and cropping intensity. Since these were in positive direction, such changes could create further demand for new technology. It was simply frustrating to note that the changes in income distribution which was found to be more unequally distributed after the withdrawal of the programme.

A very close look at the results of this study revealed that these were almost a miniature of those obtained, by many, when examining the impact of technology applied in general. The might compel one to

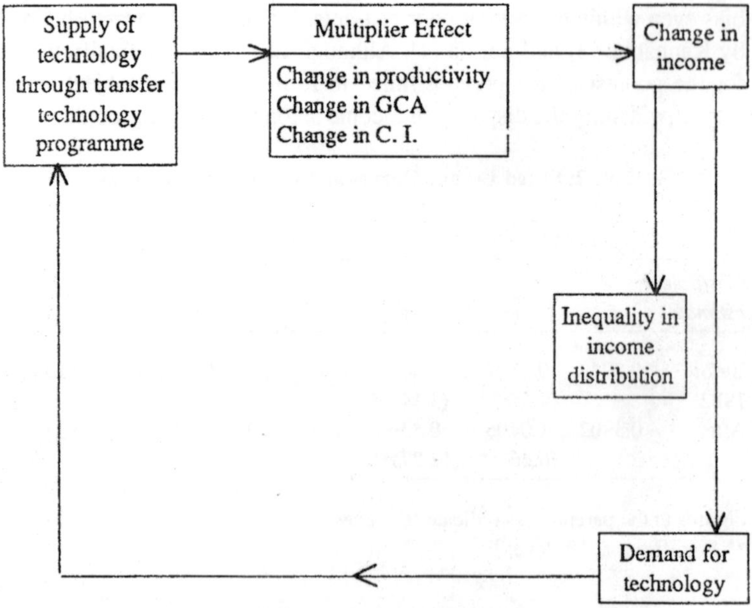

Fig. 1: Paradigm depicting various aspects of the study

generalize in the following fashion : technology never remains scale-neutral and its fruits generally follow an uneven distributions even within a group of poor peasants. If this happens to be the generalization, then one will be doubtful about whether this sorts of transfer technology programmes are necessary at all.

Transfer technology programme generally dealt with the introduction of technology without making any concomitant changes to the institutional structure. Thus it was suggested that transfer of technology programmes should either be complemented by some structural reform-measures or these measures must constitute an important component of them. This necessitates the models to be modified in reaction to the received feed-backs. Then only, it will be possible for the planners and decision-makers to bridge the gap between theory and practice—and the model building experiments will be meaningful. Otherwise, the transfer technology programmes will remain just as unachievable myths from the point of view of overall welfare.

References

Kakwani, N.C. and Poddar, N. 1976. Efficient estimation of the Lorenz curve and associated inequality measures from grouped observations. *Econometrica*, Vol. 44, No. 1, pp. 137-148.

Eltito, O. and Frigyes, E. 1968. New income inequality measures as efficient tools for casual analysis and planning. *Econometrica*, Vol. 36, No. 2, pp: 383-396.

Kannathal, V.K. 1980. Impact of Small Farmers' Development Agency on income distribution of weaker sections: A case study of Enathiri Mangalam village, South Arcot district, Tamil Nadu. *Indian Journal of Agricultural Economics*, Vol. 35, No. 4, p. 87.

Sarma, J.S. 1982. *Agricultural Policy in India : Growth with Equity*. IRDC. Ottawa, Canada.

4

Development Communication: Its Epistemological Meaning and the Experience of India

Jaya Patil and R.K. Samanta

Background

All the developing nations of the world have one thing in common that is the continuous search for a workable and viable strategy for development to redress the problems of landlessness, poverty, high growth of population, unemployment, rural-urban drift and an unequitable distribution of food and income. To overcome these problems and to face these challenges, 'Development Communication' is the only hope and answer. Development Communication is 'communication revolution'—that is using communication media and techniques to promote development and achieve desired objectives.

The term 'Communication revolution' has frequently been used in recent years to describe one aspect of changing world development. It is true that accelerating technology in the post-war period has provided Radio, TV, electro-magnetic equipments to record sound and image, instantaneous world transmission of signal through communication satellites, photo composition and high speed presses, computer storage and retrieval of information, but how much they have contributed to the development process?

The Communication revolution is a symptom of profound change as one experiences being thrust into a new century. We seem to be at a point of economic and value transition.

Communication research is as old as humanity itself and critical points in our history has transformed our lives. The potential of the new communication technology is far reaching, but the important issues relate to how we can integrate these advances in our personal, business and political lives.

Again, technological progress does not necessarily mean human progress. The tools of communication has as much power to alienate people as to bring them together. It is not just the technology of future communication that commends our attention but also willingness of people and nations to use it to enlarge understanding information marshalled continuously for the advancement of society. Flow of information has accelerated at a fantastic rate, since mankind started scratching records on clay tables or on the walls of a sheltering cave. Without communication it is impossible to bring about meaningful change and development.

Generation of communication varies from programme to programme, place to place, people to people, society to society. It rather confuses and complicates if too many development programmes are introduced at the same time. This is a hurdle in serving the needs of the people.

Increased communication facilities is both a boon and a curse. It is good to introduce new technologies but they should be monitored and evaluated till they are made use of by the beneficiaries. In areas of monitoring and evaluation a lot of care and attention is needed because huge amount is invested by Government in each of the programmes.

The role of communication is vital in mobilising people equipping them with new skills. This has been emphasised for long by Government in the five-year plans through TV, radio, films, songs and drama, field publicity department, etc., but it is only a partial fulfilment in India's 40 years of Independence. The concern is to reach information even to remote areas in form of literature or a message.

Development should take place at a faster pace for strategies to yield results for programmes like Integrated Rural Development, poverty alleviation and other developmental programmes brought out for improving the conditions of rural people by the Government.

Communication and its Essence

Communication was effective in the field only after 1960s. Prior to that Agricultural Universities did include extension education as a part of their study and made field workers as motivators, structured on the lines of American 'Farm Families'. With the launching of the Development Decade the focus was on economic issues, including agricultural news and information to drive into the heart of development issues.

The industrial revolution in Europe and the United States, the colonial experience in Asia, Africa, Latin America and the emergence of the capitalist philosophy led to the concept of development in Sociology, Economics and other Social Sciences. Wilbur Schramm was one of the pioneers in this field and his book *Mass Media and National Development* (1964) was the most influential in explaining the role of communication. In the fifties and after, industrialisation was considered the main source of development and the under-developed countries were advised to industrialise by opening mills, factories, steel plants, etc.

When Britain faced a crisis in the World War II, Churchill came forward and offered his blood, sweat and tears to safeguard the freedom of their country. When Subhas Chandra Bose formed the Indian National Army to liberate his country from Britishers, he had nothing to offer but starvation, death and misery. When Gandhi declared non-violence method on Britishers, he had nothing to offer but Swaraj, the dream of liberty. Lenin offered Socialist Society in Russia in 1917 in October revolution. This proves what motivation can do, and can be instilled in every phase of human activity. And this motivation is the essence of communication.

At one end development indicates long term material improvements in flow of goods, services, on the other it is supposed to bring changes in social, cultural and political institutions. The combined effects of the two extremes determines the level of development.

Development is the accumulation in the stock of human knowledge as well as growing capacity for learning. The relationship between development and communication is a question of how we define communication. Communication is a prerogative and right of all. But how far communication has been able to fulfil the genuine needs of people through development? Neither the medium nor the message is the realistic basis of communication. People are the cause, subject and object of development. Needs are the fundamental expression of humanity's driving force to assert its nature in dialectical tensions moving in all dimensions.

A self-reliant approach to development can succeed only if people participate. The rich resource is manpower in India but the country is unable to use it to fuller extent. The weight of experience, cultural ethos, perception of problems and solutions rarely enter into the calculus of development.

There is mistaken notion that development communication is extended only in the sphere of agriculture extension alone. It applies to all sciences and other related subjects of development too.

Development Communication

The term 'Development Communication' was first coined by UNESCO and later became more familiar for others to follow. In a symposium at the University of Philippines development communication is described as 'the art and science of human communication' applied to speedy transformation of a country and the mass of its people from poverty to a dynamic state of economic growth that makes possible greater social equality and the larger fulfilment of human potential. It is evident from this that the concept of development communication grew out of the field of agricultural communication in the same manner as agricultural communication has grown from agricultural journalism.

Development Communication is not a new concept but it's emergence and integration is gaining momentum for further scale of growth of the nation. Integration cannot be forced, it should stem from grassroots level. The important dimension of development is to establish rapport with the beneficiaries of the programmes through a purposive link with the villagers. The felt-needs of the people have to be articulated to bring about desired results. It is based on participatory democracy.

It is interesting to note that Development Communication applies in all walks of life, in all subjects and in all growths either individual, institutional or national. Development communication has been instrumental in the transition of traditional societies paving way for modern societies.

The need of the hour is to establish a direct development communication link with rural masses by the policy makers like recent introduction of Jawahar Rojgar Yojana introduced by Govt. of India in 1989 where funds flow directly to Panchayats (village local governments) rather than sanctioning it to the state governments. This takes less time in implementation process. The government has involved people and the implementing agencies at the villages, so that the panchayat officials are directly answerable to the government. In the same manner, such bold steps need to be taken by government to solve the problems of the people by involving themselves directly with the masses to feel the pulse of people.

Communication and Development are two independent as well as dependent variables from psychological point of view. Communication is one variable and development is another variable. They cannot exist without each other but latter variable bears more weightage because development and communication are part and parcel of every day life. Lack of coordination between these variables can be detrimental to the foundation of any individual or country. This is a challenge to the growth of any nation.

Realisation has dawned in the minds of scholars, researchers, academicians that any *innovations* to be implemented can be done so with the support of communication in development or vice-versa. Innovations are part of development communication process. Too much of input without desired output has affected the process of implementation. This ultimately makes any mission a failure.

Development Communication is more than a multi-media approach. It is an integration of mass media and inter-personal communication, change agents and extension agencies. Development communication should be conditioned to the required needs of the people in the manner that, it is relevant, purposive and induces or enforces them to make the development process to the stage of implementation, but not neglected at the monitoring and evaluation stage.

Development communication is more than agricultural extension education or rural communication. It does not confine only to rural development, whether it may be family planning, industrial production, agriculture, rural electrification, etc., but oriented towards developmental activities. Development communication does not deal with hot stories of detentes and disasters of coups and hijacking which have dramatic potential. The stories relate to youth problems, social justice, national integration, high-yielding varieties, rural development, emancipation of the weaker sections of the society, success stories of farmers, innovative thinking on numerous problems of agriculturists, workers, livestock keepers, etc.

Unfortunately most of the developing countries are yet to appreciate development communication support to their projects. Despite a number of seminars and conferences, the administration of developing countries feel that information and publicity should be given prominence where its impact inevitably falls on sectoral or project level activities.

Integrated approach urges a systems perspective where attention is focussed on the functional relationships which exists between different components and development programmes. In the same way integrated approach to rural development has advantage over other orientations. Communication consists of webs of significance that human beings have spun around them to penetrate into the development of the society.

Development Communication: The Indian Context

After Independence, industrialisation was given priority and in the sixties, forced industrialisation was not compatible. Agriculture was considered as important as industry by developing countries. After a decade of introducing the technology in less developed countries, it was presumed that the underdeveloped countries would become developed over a period of time. The new technology was rejected by the people and this was attributed to their traditional way of thinking, social values, norms, beliefs and practices. Development policies of the 50s and 60s paid little attention to the equality of development benefits.

It was in the seventies that the concept and the role of communication in development gained a new perspective, because the theorists, social scientists found that industrialisation and technology, economic growth and qualification were not an answer to development in Asia, Africa and Latin America. In 'small is beautiful', E.F. Schumacher attacked high technology to fit the needs of the local culture.

The strategy adopted in communication for planned development efforts: (a) to inform and educate people and decision makers on the assumptions that they are rational and will make correct decisions; (b) to involve individuals as participants in a systematic effort to alter values and behaviours patterns from less appropriate to more appropriate orientations; and (c) the appreciation of power to change behaviour without necessarily bothering to inform, educate or involve individuals in decision making.

A group of scholars, on the basis of case studies of historical conditions of four countries—India, Bangladesh, Sri Lanka, China suggest a strategy for rural development in terms of fundamental human values rather than in narrower techno-economic notions of development. Sensitising the masses through developing consciousness of their potential role in economic growth and social progress, transforming society through mobilisation of land reforms, collectivisation, building infrastructure, self-reliance, liberation of women, remoulding the elites and building institutions are some of the processes suggested as a part of rural development strategy.

Two crucial elements which play an important role in development communication are 'information for the people' and 'feedback'. It is here that the pace of progress has to be fast.

Three hundred years before the birth of Christ, Emperor Ashoka prepared his famous edicts that were set up on pillars in different parts of his empire and the inscriptions informed people about the policies of his Government to persuade them to carry certain tasks and create a goodwill amongst them. The Ashokan inscriptions were in local scripts. It is an early example of instance in practice of communicating to the people.

It is a phenomenon that emerged in modern society because of the nature of the social, political, and economic institutions that have developed in the world today and behaviour of different groups of people within society. If we try to compare the business or science of development as practised today with the type of communication represented by the inscriptions of Ashoka would be like trying to compare a sailing vessel with a modern steamship.

In India, the concept of development is closely linked with the effective functioning of mass communication, the welfare of the masses and in the implementation of Governmental policies and programmes. The vital question is whether India will continue to utilise the freedom to communicate effectively with flow of ideas across linguistic and social barriers to measure India's success with developmental communication.

Participatory communication has lot of responsibility and functions. It looks forward from local leaders for resources for need-based and to development. In turn the local leaders have vested interests due to casteism, political pressures, inner conflicts and other factors that affect the process of development.

People are the object and subject of development. They should be the motivating agency for the Government to fulfil their needs and complete the set task. Motivation should not come by force but should stem from each individual. People expect the Government to implement those demands without their contribution. The development bureaucracy should set itself the task of dedicating for the cause of developing villages. They are not committed in their purpose. There is lack of mutual trust between development bureaucracy and the people for communication gap. In such a situation the policy makers find assuming responsibility for masses's development while wanting to get out of it. Naturally 'have nots' do not have access to communication. Development bureaucracy is caught in a vicious circle.

For the success of development communication it depends on many factors like what type of programmes are brought out for the people, who are the target or to which sections of the society appealing to?, what is orientation of communication for development of the people, what technology is available. What geographical, physical and environmental characteristics in particular area covered etc?

People have since long been passive onlookers rather than willingly participate in developmental programmes. On the other hand, Government should motivate the masses, to seek education, to end exploitation and develop channels of communication between the Government and the people. Communication here acts as a catalytic agent between the two.

In fact, the development efforts produced a kind of non-fatalism and non-feudalism, in which people have become individualistic in their desires to remain dependent on the social structure. They have learnt to distrust the social structure but instead of developing an autonomous initiative have transferred their dependency to the central/federal government.

Conclusion

Communication is the essence of humanity. Through communication, individuals bridge their island of loneliness and through communication societies are formed. The ability to communicate is essential both to individual success and to group activity.

Centuries ago, Aristotle asked whether a bridge can be built between Man's political capacity and environment?. Certainly the foundation stones of Aristotle's bridge must be informative, candid, continuous reporting by Government to its citizens, to make his wants and opinions known. The two-way exchange of information and ideas can bring a mutually satisfying relationship is the function of field publicity. Democracy, more than any other system of Government depends on information and interest of the people.

Traditional thinking has severely restricted the understanding of mass communication phenomenon and the study of mass communication is particularly concerned with new communication system to create, animate and influence human societies.

The 20th century more than any other is the age of human communication. We are not quite sure what to do with abundance of information or how to make good use of the speed with which it can be transmitted yet we seek more information flow and demand it in even greater speed and we want to have transmitting instruments at our own disposal.

Communication pattern of any society is part of its total culture and it can be only understood in the context of its social organisations and institutions. Thus the heritage, caste, creed-based barriers and cultural complexities with high illiteracy percentage of the country furnish a back-drop with which the communicators and electronic media have to reckon with while working together.

Communication has now proved itself beyond dispute as an important faction for many successful projects and programmes in the spheres of agriculture, education, family planning, health and nutrition. Properly designed and coordinated with other development efforts, development communication has shown that it can inform people of new ideas and techniques, help to change attitudes and stimulate popular participation and self help more rapidly, effectively and meaningfully.

Part 2

Information Systems and Agricultural Development

"It is important to consider a broader view of access to include access to the production and distribution of messages. In a society such as the United States, where penetration for most mass media is almost 100 per cent, access quite clearly can not be defined purely in terms of physical access to the reception of mass media. In developing countries, where the mass media have not penetrated as deeply, it is natural to first consider reception questions. Nevertheless, researchers should be careful not to repeat the same top-down and passive receiver view of mass media prevalent in earlier research."

Larry Shore, 1980.

5

Information Systems in the Agricultural Development Process

Harry A. Carey and Blannie E. Bowen

Change is central to the development process, and a number of favourable conditions must exist in order for positive change to occur. Meaningful change must occur within people and be exhibited in their actions. The proper application of communications principles, knowledge and skills can enhance the exchange of new ideas and motivate people to put new ideas into practice.

But too often the communication process is thought of as a one-way process that dispenses technical information to a specified population. Achieving the objective of informing an audience on a particular point is one thing, but convincing them to change their age-old practices or invest in new technology is quite another. In some cases technological innovations are passed down from above with little consideration for, or interaction with, the target audience or those who work closely with them in the field. As a result, many of these nearsighted efforts have failed to meet the needs of the people they were designed to assist.

Some see development communications mainly as applied through mass media, such as disseminating information by newspaper, radio, or television. Mass communications techniques are often pursued as being more cost-effective than underwriting the expense of building an extension system that requires considerable manpower and monetary resources. Mass audience channels are well-suited to certain functions, such as relaying news, announcing activities, or initiating discussion on new ideas and technology, but these mass channels have severe limitations when dealing with complex issues, or at the persuasion stage of the diffusion process. In addition, mass media are not always easily accessible to poor, rural people. A UNESCO (1979) study indicates that mass media efforts alone are not adequate to meet the communication needs of Third World countries.

Another scenario would have field workers interacting with the clientele; interweaving their instruction with traditional methods and beliefs. Influential people at the local level would be consulted and appraised of the intentions of the program and encouraged to provide input into the planning process. Reputable, respected field staff would blend farm visits and meetings with instructional materials such as booklets and visual aids to inform the target audience. Field demonstrations may be used to finally show the advantages of the suggested improvements and to convince farmers to change their habits and apply the new methods on their own farms. In addition, these field activities may be coordinated with national or district newspaper or radio messages to form a broader strategy aimed at technology transfer. In this scene the development program belongs to the people of the countryside, with the sponsoring agency playing a support role. The suggested improved technology is shown to benefit the rural people directly.

Another common shortcoming found in many development communications efforts is the isolation of the information process from other activities related to the development and implementation of technical innovations. It is important that researchers and extension staff share the responsibility for the processing and dissemination of educational support materials. It must also be recognized that an aggressive communication support mechanism requires not only a firm commitment on the part of officials and relevant staff, but also adequate levels of funding to sustain an effective system.

Policies and Organizational Structures

Pandey and Bhargava (1984) emphasize the importance of recognizing the political dimensions associated with most communications functions. Too often information dissemination fails to be a link to the rural poor and those most in need are unable to access and utilize the knowledge so necessary for their existence. This is where a properly designed interactive system that links research, information and extension can make the difference that results in effective knowledge transfer.

The lack of genuine collaboration among research, information and extension personnel is one of the greatest limitations to the development process and technology transfer. Far too often these groups are viewed as diverse bodies with dissimilar purposes and they can function effectively while isolated from each other.

Research organizations are frequently administered through a different chain of command from information and extension. The role of researchers may be seen as technical experimentation and development without much responsibility for interpreting and adjusting technological innovations into useful and practical recommendations that meet the needs of the farmer. A separate organizational structure, or weak linkage, discourages researcher participation in training extension field staff, thus missing further opportunity to interact with those most attuned to the needs and concerns of the clientele. Obviously, such a structure will limit the opportunity for extension staff to learn from researchers. A similar situation is often found with universities as well, where personnel from a ministry of agriculture are discouraged from freely interacting with those in the academic institutions in the ministry of education.

Many times the information section is seen as simply a service group that is expected to broadcast or print whatever is handed to them. They are generally expected to assemble information into slick packages that everyone will want and use, but there is limited encouragement for their participation in the process of tailoring the researchers' recommendations to meet the needs of the clientele. Researchers are reluctant to become involved in producing educational materials beyond the writing of an initial recommendation and, perhaps, reviewing a production for technical accuracy. Information section personnel who are discouraged from interacting with extension field staff will be limited in their understanding and appreciation of the grassroots situation and have difficulty producing educational support materials that are effective in communication and acceptance by field staff or the clientele.

Extension field workers are expected to deliver research recommendations to farmers and convince them to accept these ideas and put them into practice. They are to somehow utilize this unadapted information in a manner that is meaningful and useful to the farmer.

Cultural Considerations

When considering where to look for credible sources of new ideas, three lines of reasoning may be pursued: (a) that educated, professional staff have most of the answers and that farmers should unquestioningly accept their advice; (b) that farmers have the ability to figure out what's best for their own farms; or (c) that the research system does produce worthwhile adaptations and innovations, but researchers can develop useful insights through listening to farmers (Lionberger and Gwin, 1982).

In order to be effective, the development process must involve those people to whom a development program is aimed. To accomplish this, it is necessary that research recommendations, educational materials produced by the information section, and extension activities reflect input from the grassroots level. Watts and Claar (1983) point out, along with many others, the necessity of incorporating social sciences with technical agriculture in agricultural development strategies.

In regard to agricultural research, the farming systems research approach has gained in popularity. Hildebrand and Poey (1985, p. 9) write, "On-farm trials are not meant to try to simulate experiment station conditions in farmers' fields. Rather, they are designed to help detect differences under typical farmer management practices and environmental conditions". This is a crucial first step in making recommendations that are appropriate for the farmer.

The task of processing information that will effectively communicate relevant research findings requires knowledge of an interaction with extension field staff and farmers. It is necessary to have an understanding of how information is passed on in a given society, and to utilize existing systems whenever possible. Educational materials have a greater chance of proving effective if a regular system for field review or testing of materials is used. With a little guidance, extension staff and/or clientele can point out weaknesses in educational materials that go unrecognized by professionals at the regional or national level.

A Team Approach

It is commonly accepted for researchers to write their reports in a manner that assures accuracy and attention to detail. Fellow scientists are the customary target audience. Since this scholarly approach is not readily understood by extension field staff and their clientele, the information interpretation and processing function becomes critical (FAO, 1984). Research findings must be translated into understandable

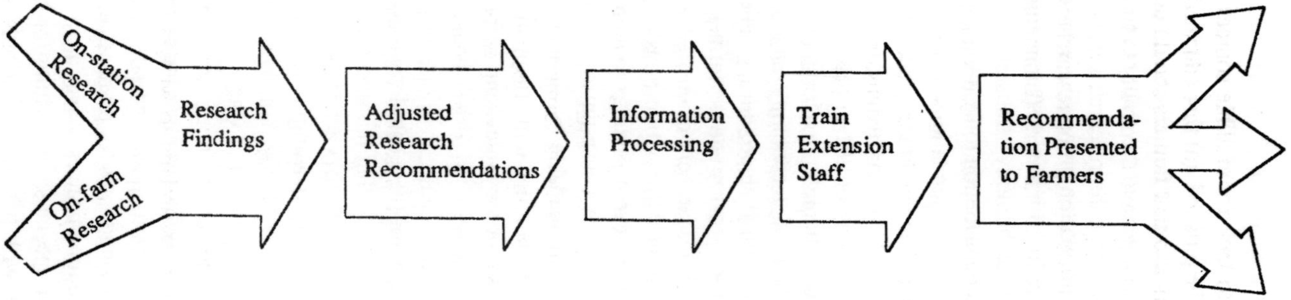

Fig.1: On-station and on-farm research findings are converted into technical recommendations that reflect the social and agro-climatic conditions and constraints faced by the farmer. This information is then packaged into educational materials that are understandable and useful to extension field staff. After appropriate training, field staff present the recommendations to farmers.

messages that have practical meaning to those who are to put it into practice. Former chairman of the Board for International Food and Agricultural Development, Clifford Wharton, Jr. (1983, p. 11) states; "We know how to harness the creative and inventive forces of science and technology, but I submit that we still have not been fully successful in technology diffusion".

An effective agricultural information processing system must sustain a deliberate and determined effort to join research, information, and extension personnel in the technology transfer endeavour. This interactive system encourages contribution toward the common purpose of developing and producing recommendations and materials that are understandable, meaningful, and relevant to the needs of farmers.

1. The current research climate has an increasing consideration for adapting recommendations in view of the social and agro-climatic constraints faced by the intended clientele. In addition, researchers must be concerned with the transfer of technology and assist with the preparation of suitable information that provides practical and useful recommendations to the extension worker and the farmer.

With this objective in mind, researchers often find that they enjoy applying their creativity toward the information transfer process. They find new challenges in adapting research findings to more practical and useful suggestions that recognize the constraints confronting farmers, and in writing manuscripts that are readable by the non-scientist. Some find reward in providing ideas for field support guides with communicative charts and illustrations, in devising helpful visual aids, or in inventing a workshop exercise to drive home an important technical point with extension field staff. In short, they buy into and accept a measure of responsibility for the technology transfer process.

2. Communications staff must see that the intent of research is put forth in such a way that insures reasonable accuracy, while converting the information into understandable, meaningful messages as deemed appropriate by evaluative feedback. Ideally, communicators would be regular members of program planning bodies where they assure that the resource allocation and lead time for communications needs are duly considered at an early stage. It may be desirable for communicators to share with extension training specialists the leadership in organizing a system for selecting and prioritizing topics for production.

It is often useful for communicators have a role in training researchers and subject matter specialists in writing for lay audiences. An investment at this stage will pay rich dividends as the initial manuscripts will arrive in the information section in much better shape. Upon receipt of the original document, the information staff begins the editing process, aiming at improving the readability for the intended audience. This may require a significant effort in working with research and extension personnel in simplifying the writing and developing creative approaches that provide visual support to graphically demonstrate the message. This support may be in the form of graphic innovations within a field support guide, separate visual teaching aids, or an in-class training exercise. As researchers become more skilled at inputting into the information processing system, they provide many helpful ideas and make the task of the information specialist more enjoyable.

As encouraged by Evans (1985), communications staff need more than just traditional information skills associated with producing and distributing materials. Communications specialists share with extension staff the concern of determining how farmers obtain information, what communications channels are useful, when to use them, and how to relate educational support materials to the ongoing extension program. Additionally, they must evaluate the effectiveness of the materials produced. Communications personnel can also play a key role in providing extension staff with training in communications skills. The importance of extension methods and communications training is pointed out in a survey by Sigman and Swanson (1984) of extension administrators in 129 third world countries. The study showed the lack of pre-service and in-service training in extension methods and communications principles and skills. This area ranked second only to the extension worker mobility problem.

3. Extension staff must assist with identifying subject matter topics where information support is most needed, and help in generating appropriate support materials. They must work with communications staff during the field testing and the feedback process to ensure the end product is practical and useful. Extension staff must also assist in making necessary changes or in generating solutions to recognized problems. Extension personnel should take the leadership in determining field staff training needs, and in providing training programs to meet those needs. They can plan and coordinate the necessary technical instruction along with training to help field staff appropriately apply the information. Extension inservice training is often regarded as the most important means by which to upgrade the effectiveness of field staff.

The tasks of researchers, subject matter specialists, communications specialists, and extension staff interrelate extensively in most effective information processing systems. Effective educational materials are difficult, if not impossible, to produce with researchers, information and extension personnel working in isolation from one another. It is often stimulating and enjoyable for the researcher to assist in developing a useful teaching aid; for the information specialist to help train field staff; or for the extension worker to contribute to the production of a field support guide. This interweaving of duty imparts a better understanding and appreciation for the entire process to all participants and encourages a more effective team approach.

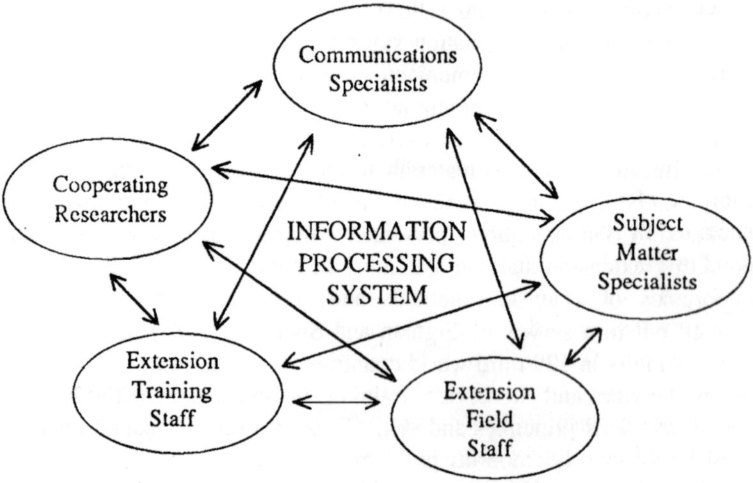

Fig.2: An effective information processing system involves information specialists, researchers, subject matter specialists, extension training staff, extension field staff and others in a variety of interrelated activities aimed at producing understandable, practical educational support materials.

Cooperation can Solve Problems: An Example

The illustration presented here shows how a team approach to educational materials production can identify potential problems and find ways to overcome them. This example begins at a time when on-station and on-farm agricultural research began producing early recommendations related to increasing maize production in Swaziland, southern Africa.[1]

1. Research indicated that, based on soil tests, fields to be planted to maize should have certain basal fertilizer and nitrogen topdress requirements. Recommendations were given in kilograms of fertilizer per hectare.

2. Information staff, working with researchers and extension staff, produced an easily-read, understandable publication that explained the recommendations. Charts and tables converted fertilizer recommendations into fertilizer rates related to the locally-available fertilizers. This was a process requiring considerable interaction and compromise to move from basic research findings to simplified, understandable recommendations for use in the field situation.

3. Information and extension training staff tested draft copies of the fertilizer recommendation publication with extension field staff. The field staff pointed out the difficulty they would have in utilizing any recommendations given in kilograms per hectare since they had no scales by which to weigh the fertilizer. A further problem was the shortage of measuring wheels with which to measure the field size in hectares. In addition, much of the fertilizer was commonly applied by hand in a band along the row.

4. In order to overcome the problem of how to measure fields in hectares, researchers, information and extension personnel created a field support guide. The guide explained how to determine an individual's pace coefficient and how to measure fields of various sizes and shapes by pacing. This pacing and measuring publication was then field tested and appropriate suggestions from field staff incorporated into the final content of the guide. Practical training in pace coefficient determination and field measurement was provided to all extension field staff.

5. The obstacle of field staff not having scales to measure the fertilizer was overcome by researchers, information, and extension personnel again working together. Fertilization recommendations were developed based upon the 350 ml. cold drink can, which is readily available throughout the country. A publication on the cold drink can method of fertilization was developed and field tested. This field support guide contains explanations plus recommendations in the form of charts that show the distance along a row one cold drink can of fertilizer should be spread to fulfil a given fertilizer recommendation. Again, training in the use of this publication and how to accurately apply fertilizer by this method was provided to the field staff.

In this example, the first publication was easily-read and understandable, but it failed to meet the needs of the target audience. It was all too typical of how the production of educational materials in the context of western culture can become a fruitless effort. The ensuing team effort was efficient in identifying and overcoming the problems and eventually produced effective educational materials to support the extension field worker in implementing the recommendations.

Information Processing

Recognizing the limitations of utilizing mass media, a number of experienced agricultural development specialists suggest that printed material is likely the most important medium for supporting an effective extension system in developing countries. Field support guides or fact sheets can:

a. help assure the accuracy and consistency of research recommendations and/or information that is being desseminated to field staff, farmers and others;
b. support and supplement extension inservice training as well as serve as a working reference to extension field staff over a period of time;

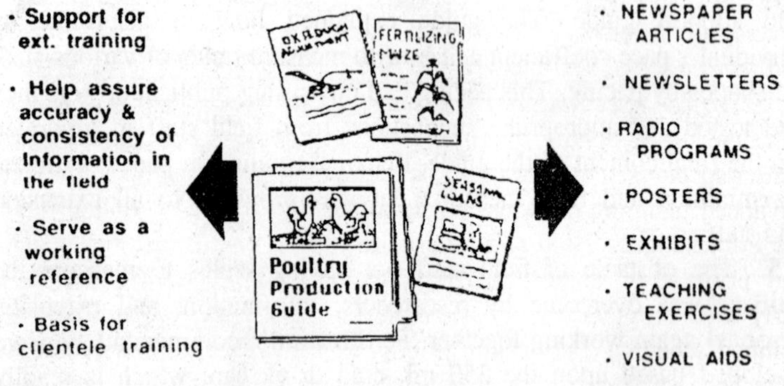

Fig. 3: Field guides are useful in a variety of ways to support extension field staff provide baseline information used for generating other educational support materials.

c. serve as instructional material or a basis for information used by field staff for specific farmer training programs; and
d. function as a reservoir to spin-off information for other media efforts such as radio programs, newspaper articles, newsletter features, posters, displays, and visual teaching aids.

It is important for every extension organization to create an agenda, and a system to keep the agenda current. There should be an ongoing process to identify and adjust the organizational objectives. In many developing countries this includes the identification and prioritization of subject messages in order that the information processing system produces the most important educational support materials first, or produces support materials in a timely way to coincide with field instruction.

Operation of the Information Processing System

This is a typical method of operation of an information system for producing extension support materials in a developing country:

a. *Audience:* Materials are aimed primarily at the extension field staff, with the farmer as the final target clientele. Certain farmers who can ready may use these materials directly, or have a family member or neighbour read the materials to them. Other audiences may include: national subject-matter specialists, research staff, other rural development field staff, personnel in other ministries, school teachers, and private sector agri-business people.

b. *Subject Selection*: Needed subject topics are suggested by farmers to extension field staff, by other extension officers, by national subject matter specialists, research staff, and others. An increasing emphasis on advanced planning of extension messages and a backlog of projects in the information section require a more formal process for determining and prioritizing support materials production. Identifying the topics early will provide the opportunity to generate these materials on a planned schedule. Publication priority is usually given to subjects that have been selected for extension messages, which are generally those relating to the technologies of greatest potential for improving agricultural productivity.

c. *Authorship:* After identifying the needed subject matter topic, one or more authors are identified as having responsibility for producing the initial manuscript. Researchers, extension subject matter specialists and others are encouraged to work cooperatively in developing this information. Contributors may be found beyond the ministry of agriculture's research and extension systems, with authorships from other agriculturally-related project staff, university faculty, and agriculturists from a variety of other projects. The use of specialists from the private sector may be helpful, such as bank officials contributing to credit publications or equipment dealers supplying information for an oxplow leaflet. The solicitation and incorporation of relevant information from other countries is also encouraged.

The information staff, along with extension staff and subject matter reviewers, assist authors by providing direction in keeping the manuscript focused on a reasonably narrow topic that fits with other publications that are either planned or have been produced. Consultation and/or training in writing for non-scientific audiences is frequently provided to authors. They may be given examples of readable, low-tech writing that makes use of tables, charts and illustrations, with the aim of generating information that is clearly understood by the intended audience. Related technical subject information should be gathered or provided to the author. After following this interactive process of producing a practical extension support guide once or twice, authors usually improve their writing dramatically and input better manuscripts into the information processing system.

In this age of enhanced word processing, it makes increasing sense to be concerned about authors' accessibility to computers that are compatible with those in the information system. In addition to the information section, computers may be made available at the research station, the location of subject matter specialists, and the extension training section. It may make sense to have one or two additional "loaners" available to those who may input documents and are not located for convenient computer use.

d. *Editing, Graphics, and Layout:* Some manuscripts received as hard copy and are entered into the computer by a secretary. Others are received as pre-entered documents on a computer disc, reducing the demand upon the information section's secretarial staff. One or more editors then review the information on the computer, making editorial corrections in sentence structure and grammar, along with formatting and organizational suggestions and changes that enhance the readability of the information. These changes are based on proven principles of increasing the readability and understandability of writing through word size, sentence length, line and paragraph length, logical organization

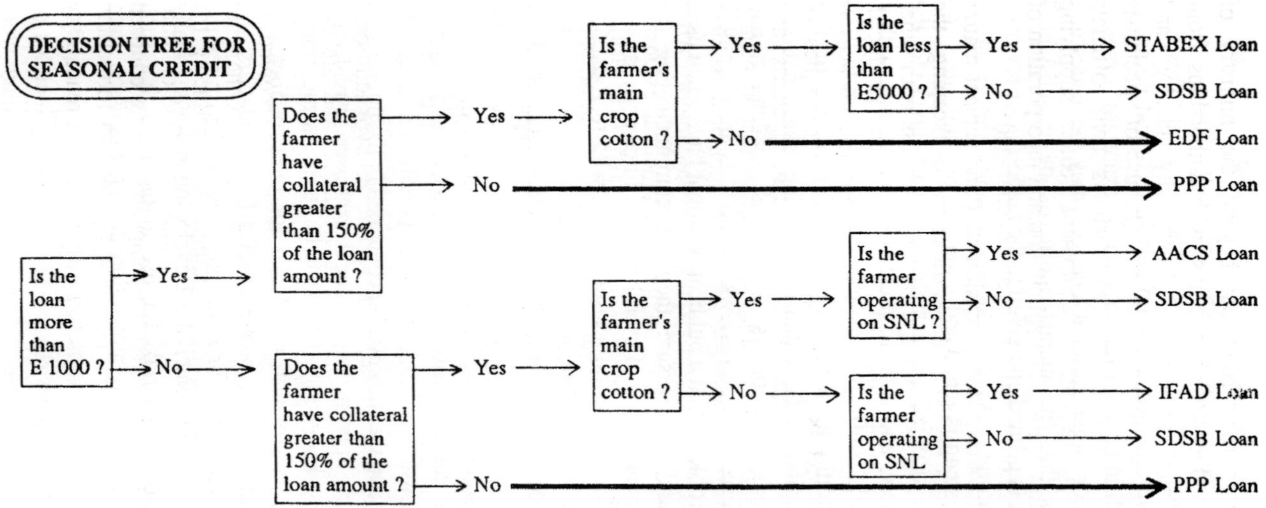

STABEX Loan -	European Economic Commission funds to assist cotton farmers with drought problems.
SDSB Loan -	Swaziland Development Savings Bank loans for small to medium input farms.
EDF Loan -	European Development Funds available for medium to large loans on non-cotton farms.
PPP Loans -	Peoples Participation Project funds for low-interest loans to groups of farmers.
AACS Loans -	Agricultural Advisory Credit Scheme for cotton farmers on Swazi Nation Land.
IFAD Loans -	International Fund for Agr. Development for marketing, credit, and irrigation schemes.

Fig. 4. Visual approaches can be invaluable in conveying information. This flow chart is a centerfold in a ten-page booklet on obtaining seasonal loans. It assists field loan officers to quickly and accurately determine the proper loan to apply to a given situation.

and flow, use of titles and subtitles, letter style, size and clarity, the use of illustrations, white space, layout, and other graphic treatment.

Information staff often find it necessary to check several times with the author(s) to approve changes, clear up certain technical points or have them provide additional information, photos, examples, or other support materials. Major organizational changes may be necessary. When the editor feels that the manuscript needs more substantial alteration, a rough draft is printed and given to the author(s) for rework.

e. *Review and Testing*: As the document nears the point of completion, a layout/pasteup is made and a number of "dummies" are constructed. These photocopied publications are then given to the author(s) for final review, and used to pretest the publication with a group of extension field staff. Information intended to be utilized directly by farmers should be tested with typical members of this audience. The information staff, extension training staff, national subject-matter specialists, and others assist with evaluating and pretesting of these materials. These reviews and pretests are extremely helpful in identifying problems that a publication might have when used in a field situation, and help assure they are communicating the information in an understandable and useful manner. Final copy is then typeset in the information section on a laser printer, with reproduction on an offset press. (It is important not to overlook the review and possible contribution of all potential subject specialists, or these individuals may not feel a part of the program and withhold their support for the materials after production and distribution.)

f. *Utilization:* The initial distribution of extension field support guides is linked to extension training whenever possible. Appropriate background knowledge and the information contained in the publication are explained and, when appropriate, practical exercises provided to assure a proper working knowledge by the field staff. As mentioned earlier, the educational information developed for these publications provides a basis for spinning off useful articles for newsletters and newspapers, serves as a resource for radio programs, and inspires the development of visual support materials. Developing nations usually possess so few educational materials for use in the field that such materials increase enthusiasm and motivation among extension field staff.

Recommendations

An effective information processing system can produce educational materials that are of considerable value in assisting field staff in their jobs. Appropriate support materials can enhance the exchange of new ideas and help motivate people in putting new ideas into practice. The following are suggestions for developing an effective information processing system:

1. Develop a systematic approach for selecting and prioritizing educational materials production projects that reflect the needs of the extension field staff and farmers.

2. Seek a broad base for obtaining relevant information and authors to produce educational materials on the selected topic. This will reduce the likelihood of overlooking important facts, of wasting resources on reinventing the wheel, and increase the numbers of those who "buy" into each project, thus insuring the acceptance of materials that are produced.

3. An information process must be based on mutual commitment by researchers, subject matter specialists, communications specialists, extension training personnel and extension field staff. It's generally best to have the production of educational materials written into job descriptions. Staff should be selected with these interests and skills right up front.

4. There must be recognition and appreciation by all players of the necessity for the process to interactive, with useful suggestions accepted from a broad base, particularly including those in the field.

5. All participants should feel an obligation to produce educational materials containing recommendations that are applicable to the intended audience, considering their resources and their social and agro-climatic conditions and constraints.

6. A conscious effort must be put forth by all to produce educational materials that are readable and understandable to the audience(s) for whom they were produced. In addition to writing in a common, understandable language, these materials should maximize communications by using visual materials such as illustrations, charts, graphics and photographs.

7. A field review and feedback procedure can be invaluable in assuring that the materials produced are effective, both in content and presentation, in meeting the needs of the clientele. Upon recognizing the merits of an effective and growing information system, researchers, national subject-matter specialists, and others will increase their

commitment and support. They will generate a flow of useful and practical information to be processed and packaged for use by the extension worker to assist the farmer in improving productivity. Efficient and effective information transfer can best be accomplished through a combined effort where everyone understands and contributes to the process. Such a process should encourage information flow in both directions to insure the usefulness of the information.

Notes

1. The Swaziland Cropping Systems Research and Extension Training Project (CSRET) is a USAID-funded project with The Pennsylvania State University, Tennessee State University and the Government of Swaziland cooperating. Carey served as on-site information advisor from September, 1985 through August, 1987. With contract extensions, the project will continue until 1991.

References

Evans, J. (1985). International communications education and training: A look to the future. *Proceedings of the International Agricultural Programs and Agricultural Communications Conference*, St. Louis, Missouri, February 13-15.

FAO. (1984). *Agricultural extension: A reference manual*. United Nations Food and Agricultural Organization, Rome.

Hildebrand, P.E., & Poey, F. (1985). *On-farm agronomic trials in farming systems research and extension*, p. 149. Lynne Rienner Publishers, Inc., Boulder, Colorado.

Lionberger, H.F. & Gwin, P.H. (1982). *Communications strategies: a guide for agricultural change agents*, pp. 209-210. Interstate Printers and Publishers, Danville, Illinois.

Pandey, S.R., & Bhargava, A. (1984). Alternate development communications efforts for science education in rural India. *Proceedings of the Midwest Regional Symposium on Development Communications in the Third World*, the University of Illinois at Urbana-Champaign, April 15.

Ray, H.E. (1985). *Incorporating communications strategies into agricultural development programs, parts 1-3*. Academy for Educational Development and the Institute for International Research.

Sigman, V.A., & Swanson, B.E. (1984). Problems facing national extension in developing countries. *International Programs for Agricultural Knowledge Systems (INTERPAKS)*, University of Illinois at Urbana-Champaign, May, 1984.

UNESCO (1979). *Low-cost communications systems for educational purposes in third world countries*. Paris: Methods, Materials, Structures and Techniques Division.

Watts, L.H., & Claar, J.B. (1983). Knowledge transfer for agricultural development: history, limitations, importance. *Knowledge Transfer in Developing Countries*, the proceedings of a Conference on International Extension, Steamboat Springs, Colorado, July, 1983.

Wharton, C. Jr. (1983). *BIFAD'S sixth birthday: A personal exaugural*. BIFAD Briefs, 6, 4, Appendix pp. 1-11.

6

Dimensions of the Microrevolution in Learning

D.R. Minnick and E.V. Tipa

When societies pay accolades to the minstrels, clowns and gladiators, the real movers and shakers of the world are those men of science whose inventions bombard our lives. One such invention is the computer. The implications of the computer to human information processing, banking, labor markets, mechanization, transportation, education and every conceivable aspect of our present and future environment is awesome.

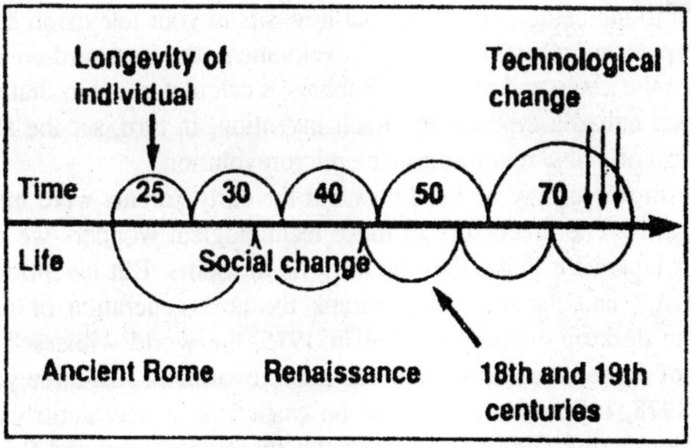

Fig. 1

Adapted from: M. Knowles. 1972, The Modern Practice in Adult Education: Andragogy vs. Pedagogy, New York Assn. Press.

The Birth of the Microrevolution

To understand the implications and impact of this entity, the microcomputer, and its associated revolution on societies, it is useful to have a perspective of the transitions that the human populations have undergone through time (Deevey, 1960; Toffler, 1980).

For survival, man the animal developed some necessary components in the dawn of civilization, such as the taming of fire and the development of a communication tool— a language. Languages are, of course, a precursor to the array of low—and high-level computer languages that are assisting us to redefine communication; a new term has been inserted into our language — computer literacy!

By about 8000 B.C., man had progressed from the cultural to the agricultural revolution by establishing a stationary rather than a nomadic food base.

The pace of technical discovery quickened in 1650 A.D. with the advent of the scientific industrial revolution. It is in this revolution that the additional pinions of the microrevolution were invented. Water power was eventually harnessed to generate that new kind of power, electricity. Electricity was pumped down wires and powered a new communication device called a telephone. Electrical impulses were again pushed into the airwaves and radio emerged as the electricity conducted into the set lit up its composite of vacuum tubes. The vacuum tube succumbed to the transistor as still another new technology evolved. But the tube technology produced radar tubes which led to the cathode ray tube that now sits in your television set or your computer monitors. In another development, the keyboard concept went from the piano and abacus to Babbage's calculator and to that new contrivance called a typewriter. Each invention, in turn, set the stage for the birth of a new revolution—the microrevolution.

The term micro may be used now but the early parents were macro in all regards. The precursors to those technological wonders we now set in our laps were huge; they literally filled rooms. But invention of the transistor and the microchip shrank the next generation of these giants into desktop entities. So that in 1975, the world witnessed the entrance of the microcomputer and the microrevolution. Just three years later in 1978, information began to be carried on a new contrivance called a floppy disk (Extension Committee on Organization and Policy, 1982). Our world had entered a new age.

The Other Revolutions

As information progressed along Deevey's time line, there has been a spin off of information in quantum leaps to produce a spectrum of revolutions. Four of these are closely integrated with the microrevolution as they pertain to learning and warrant inclusion in the discussion. They are the frontiers of space, medicine, psychology of learning and, as always, the avenue of breakthroughs in physical science that has led to advanced communication systems.

Space research has helped us understand planets and astral bodies beyond our own. It has sent men to the moon, and probes to investigate the rings of Saturn and beyond. Science demanded and got miniaturization, solar energy packs and computer extraordinaire. The spin-offs are evident today in the rapid development of the microcomputer as a learning tool. Space programs deployed satellite communication networks that allow us to exchange information with expanding populations in shortened turn-around time (Tietjen, 1987). When this information exchange by satellite occurs between computers that can store and recall large bodies of information, the implication to the learning process should be evident.

In medicine, frontier work in brain research is providing insights as to how we learn; that there is division of labor among parts of the brain —a central processing unit, storage and recall systems not unlike a computer (Brain Mind, 1977). New findings on brain function in the field of neurochemistry are making the production of "smart pills" a possibility (Fisher, 1964). Again, the implications to learning should be evident.

Cognitive psychology or the psychology of learning is utilizing the findings of brain research and advanced hypotheses of hierarchical relationships, semantic nets and memory levels of inputting, storing, and recalling information (Minnick, 1989a). The linking of these two sciences has produced a new discipline called neuropsychology (Herman, 1988). Researchers in knowledge engineering are documenting the steps leading to a science called knowledge elicitation (Levine, *et al.*, 1986). This science seeks to probe the parameters of problem solving by tracing the thought process of experts. For to what avail is stored knowledge if it is not recalled to solve problems? Research contrasting lateral and vertical thinking are substantiating the concept of right- and left-brain messages and their implication to the learning process (De Bono, 1970).

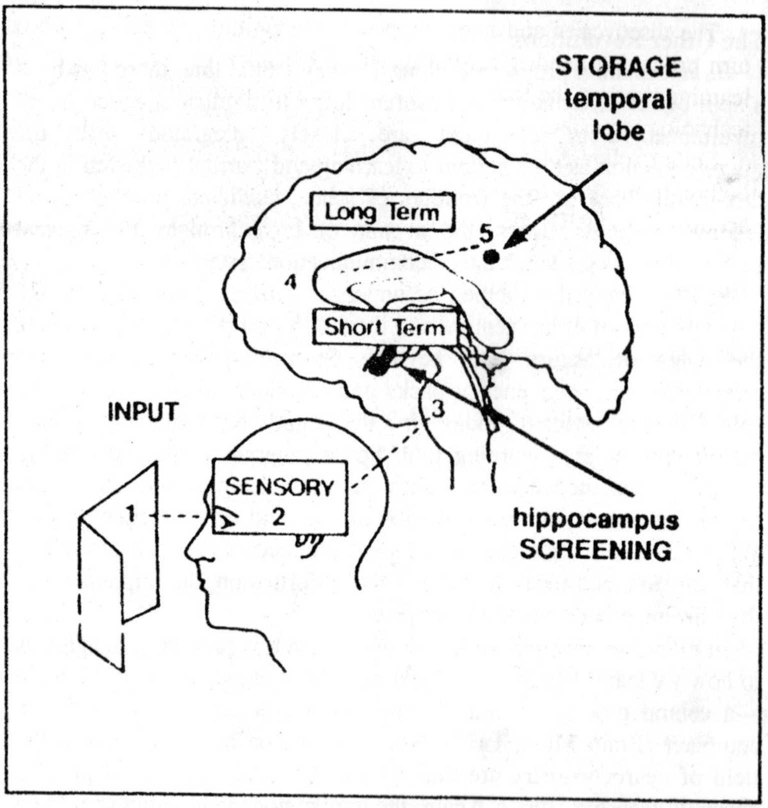

Fig. 2: Memory Levels

Finally, the frontier of *physical* and eventually *electrical engineering* is producing the basic advances that could be stored inside the computer, ready for use. It has produced the peripherals linked to the computer as well as the software that enhance the capabilities of the computer system: scanners that can throw images from paper to the computer screen in 64,000 colors at the touch and click of the mouse peripheral; resolutions of the CRT that equal those of a slide film; hand-held television cameras that can capture and relay information to the computer for the learning process, serving as eyes for the computer to allow it to communicate in our visual world; discoveries that allow us to hold thousands of pages of information in a hand held disk, and of course, the voice peripherals that allow oral communication with the computer as well as the keyboard and touch panel (Goren, 1989; Bien, 1988; Podomore and Fagug, 1986).

The discoveries and integration of these frontiers of science are in turn producing another revolution: the application of computers to the learning process, the training process, human resource development and technology transfer.

Let's explore a brief history of independent knowledge acquisition before turning to the computer per se, for independent knowledge acquisition is the forte of the computer in the learning process.

Independent Self-learning Theories

The most important category of self-learning comes to us at a very early age and stays with us through life—discovery learning. We don't need another individual to tell us that fire is hot and ice is cold if we have learned of these by discovery. One group of advocates of this learning strategy is that of the educational psychologist Bruner (Owen, *et al.*, 1981).

Examples of discovery learning are case studies and guided design methodologies wherein learners are given parameters of a problem and are guided to deduce solutions through interpolations from models or examples that they explore (Wales and Stager, 1977). Exploratory or discovery learning has a major drawback in that it requires rich learning environments and does not lend well to stringent time parameters associated with formal learning environments.

However, most learning experiences are derived from knowledge stored on some type of medium such as print on paper, slides, video tape or disk, audio tape or records, CD ROM and, of course, the floppy disk. No matter which medium is used, it is the *way* the message is constructed that is important to the self-learning process.

Now, enter the behavioral school of psychologists led by B.F. Skinner who gave us the tenets of programmed instruction, and Mager and Bloom who introduced the principles of performance objectives into the learning process. Moreover, there were Ausubel, Novak and other cognitive psychologists who provided the concepts gave us subsumers and concept maps. And on goes the list of researchers, hypotheses, concepts, principles and advocates (Knowles, 1984; Mager, 1962; Novak, 1977).

Plato

These self-learning principles were first integrated with the computer to any large extent when the National Science Foundation of the United States awarded a grant to the University of Southern Illinois to develop a comprehensive learning system on a mainframe computer. What emerged with the help of a contractor, Control Data Corporation, was an authoring system designed explicitly for developing self-learning materials or courseware. The authoring language was called Tutor and the system was named Plato.

The Plato hardware terminal was a plasma screen with a touch panel. Response and communication were dependent on keyboard strokes. This peripheral unit was hardwired to the mainframe, which could be several hundred miles away, by a dedicated telephone line. In essence, it was a subscriber system. Although some programs were better than others, there was a quality standard and the principles of self instruction were applied. Most important of all, the programs were highly *interactive* and provided feedback.

These aspects were important to the emerging technology because by contrast, the television monitors were being taken down from the ceilings of university classrooms at that time. The technology of using television to disseminate the passive lecture to large number of students had failed because of its lack of personal interaction.

So in the late sixties, one self-instruction medium was dying while another was being born. But even the Plato concept was having a difficult time in being accepted, primarily because of the high cost factor and the resistance of conservative faculty in favor of the old standby, the lecture.

The Micro

But with the birth of the micro the whole concept of self-learning was changed. The umbilical cord to the mainframe was cut. The event, slowly at first and then at a quicker pace, produced a multiplicity of hardware and software packages designed explicitly for the authoring of self-learning courseware. Information could now be carried on a disk independently, and used at independent times without the need of a learning center and a mainframe. It expanded learning environments to almost anywhere where there was electricity.

Floppy Courseware

Nevertheless, the initial years were disastrous. About anyone who could master an authoring system became an overnight courseware designer. Many of these designers who might have quasi mastered an authoring system were devoid of the principles of self-learning that could make the courseware successful in imparting information. All manner of software packages came on the market, most of which were dismal failures to teachers and learners who tried to be innovative in using the new technology

Subsequently, quality control review panels are now springing up in the disciplines to set standards and monitor quality through peer review (Note for example reviews of software packages that appear in publications like the *NACTA Journal*). With peer review, courseware packages have improved, and are even declared as creative works in academic circles. However, the paucity of quality courseware is still a factor that limits the emergence of interactive computer aided-instruction (CAI).

Features/Capabilities of CAI Authoring Systems

Feature/Capability	*Function*
CAI Development:	
Menu-driven editor	shorten development time
Language capability	tailor CAI to instructional needs
Screen Displays:	
Multiple display board support	design CAI with a variety of screen type options
Graphics	illustrate text and show relationships
Font types, sizes	display content information
Special character creation	display special fonts or characters
Color	emphasize graphics, text display
Interactive video	show real world situations with overlay of text and graphics on video
Interaction/Branching:	
Interactive	reinforce information through questions, evaluation of learner's response and feedback
Response analysis	evaluate and judge learner's response
Feedback	provide feedback on correct, wrong responses
Branching	design effective interactive frames and allow learners to choose sequence of instruction
Validation	manage a database of learner input
Special Effects:	
Audio	provide audio using tone generator or video disk
Animation of text, graphics	gain learner's attention
Text highlighting	emphasize text

The Giants

In the initial stages of self learning development on the micro, the giant corporations that were producing the hardware were not much interested. Business, big business was with business, not school systems.

It is said that in our era, the lead time for technology development is about ten years and that it takes another ten years for that technology to either die or be replaced by a new technology (Extension Committee on Organization and Policy, 1982). Self learning through computers is just now emerging as a dynamic technology in the learning process.

The heads of the corporate giants are turning their heads to look at the training and education market that prevails in the corporate, military, government and domestic school environments. The race is on, not only between corporations but nations to capture a share of the market. Some nations have cloned the technology and plummeted prices so that the strategy can now be a reality to an ever expanding population. Corporations are now donating units to the schools and training institutions around the world (IBM, 1985; Hewlett-Packard, 1986). They are assisting in designing new authoring systems and providing grants for the development of software packages. The result is that powerful new learning packages are being developed for the self learning mode. If knowledge is power, then power is available in the form of a floppy or five-inch compact disk.

Domains of CAI

With this background in mind, let us now look at some of the parameters of the different domains of CAI (Minnick, 1988). The number of these domains depends on whether you wish to clump or split information but for convenience, we will address five. First is the tutorial wherein the computer guides the learner through the information on a one-to-one basis. This is usually a passive process used for inputting information. The frames of information are presented in a programmed sequence where the computer is used much like a moving book.

The *Q and A* or question and answer, sometimes called the response, is an active process. The learner is asked to respond to questions about some unit or concept of information that has been previously inputted via the tutorial mode. Q & A is used to make the learner process the information so it is stored in long-term memory.

ICAI or the *Interactive Computer Aided Instructional* mode is where the tutorial and the response sections are integrated. It is a highly active domain wherein information is inputted in small packages and then processed in response frames. The two domains are also interactive in that repetition loops may be in the program to take the learner back over the information in the tutorial frames that were missed during the response section. It may be in a linear sequence or branched, wherein the learner is taken down different tracks of information packages depending on the answer to the questions.

The feedback can be graded into positive and negative reinforcement responses. The responses can be further graded by descriptor words used in the response. Words that denote excellent, good, fair, poor, bad and terrible or miserable performance can be used. Cues can be flashing lights and sound in addition to word responses. Previous learner responses may be stored in the computer memory and brought out at a later time to relate and give immediate feedback to previous questions related to the present information. This integration of responses to information along with the appropriate feedback cues requires a high degree of creativity on the part of the courseware designer. It also requires that the programmer have insight to the content material of the module.

Simulation is another domain of CAI. It is a highly interactive domain in which the learner inputs information parameters which simulate real-world conditions. The program is usually designed with a mathematical model of interrelated variables. The learner gets feedback on their inputs when the program relates the results of the decisions in picking the parameters.

The advantage of this domain is that it takes the learner a step further into the threshold of problem solving by getting the "feel" of the results of trial-and-error decisions. It is the threshold of right-brain learning in that it calls for intuitive and heuristic as well as analytical decisions. Another advantage is that learning can take place with shortened turn-around times frames. In agriculture, the thrust is in the area of crop growth simulation cycles in relation to physical, biological, economic and cultural factors (Whittaker and Thieme, 1989).

Finally, the emerging edge of the microrevolution that embraces learning is in the area of *problem solving*. One area of problem solving that is getting priority attention is that of Artificial Intelligence, especially the Expert Systems components of AI. This is where knowledge engineers endeavour to elicit the thought processes of the

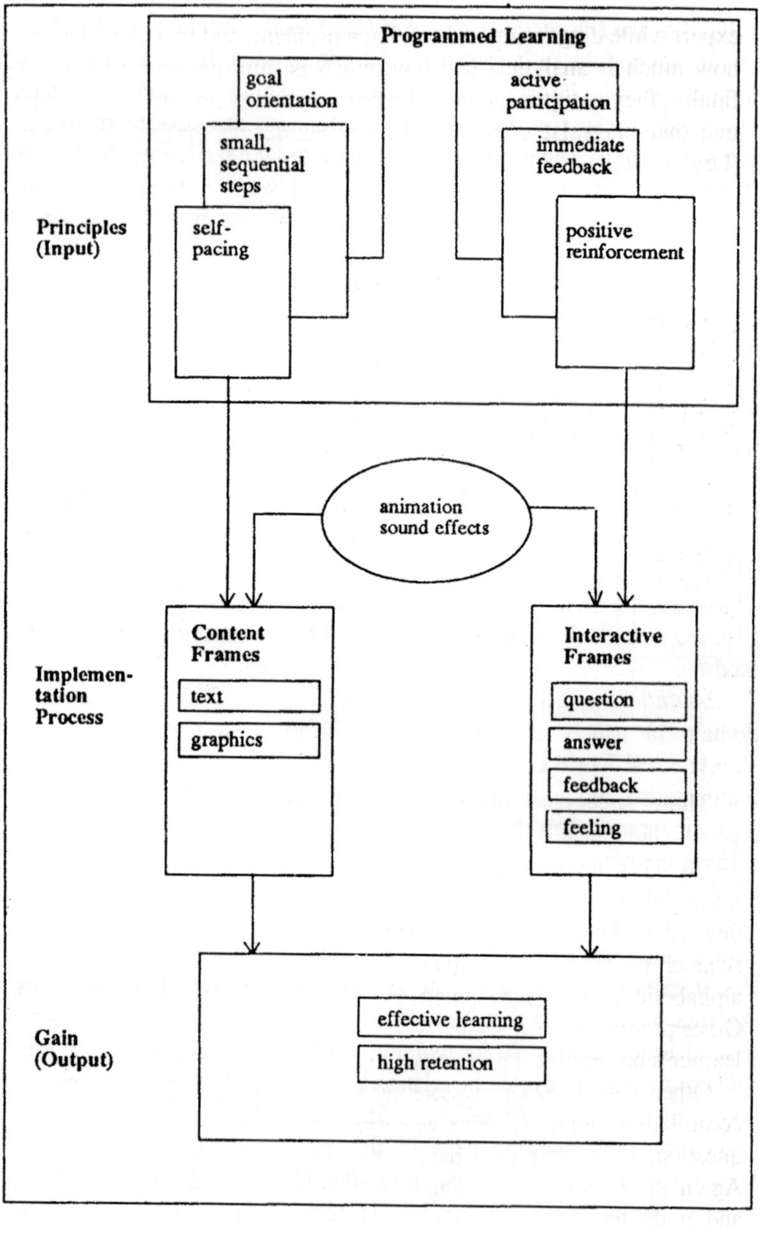

Fig. 3: CAI Model

expert while diagnosing and solving a problem, and then they determine how much is analytical and how much is intuitive or heuristics. And finally, they assign weights or probability to statements in the decision tree that interrelates the thought process during the problem solving (Levine, *et al.*, 1986). More on this later.

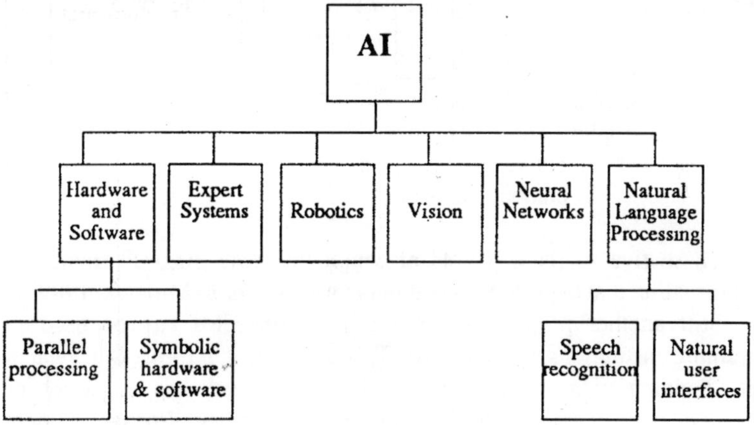

Fig.4: Applications of Artificial Intelligence:

CMI

There is another function of the microcomputer in the learning process and that is *computer-managed instruction* or CMI (Minnick, 1988). This is where the computer is used as a tool to store and recall information associated with training programs. Data banks can be devised to keep test records after they are scored with an electronic scanner. Programs are designed to recall the scores by assigned number, alphabetical order of students, ascending or descending order of scores. Other programs are designed to do updated statistical analysis on each learner's accumulated total of progressive scores.

Other data banks can store test questions for random recall and compilation into a test. This way, no two tests are alike in sequence or question, if the test item bank has a sufficient number of questions. Again the computer can remember the correct responses to each test and grade them at completion.

Programs have been designed to analyze tests as to their reliability and difficulty so that a semblance of standardization can be reached with the learner population. Each item is analyzed and a weight of difficulty assigned. Eventually the entire test can be weighted (Minnick, 1984). One CMI project that the authors have just completed is called a Knowledge Generator. It is intended to assist trainers in generating a curriculum matrix from existing courseware. It is also designed as a repository of knowledge that has been captured from the content experts and stored in *learner formatted databases.* It consists of a series of shells tucked inside each other consisting of

1 courses in the institute
  2 disciplines in the course
    3 topics in the disciplines
      4 objectives in the topics
        5 content frames that define the objective

These five shells will hold at completion over 300,000 frames of information that capture the content of seven courses (Minnick, 1989b).

Still another project that exemplifies how CMI can be used in training institutions is called MIST, Management Information System in Training. This project, which is in progress, has an array of components to manage training programs including: finance, student records, personnel information; search systems of training colleagues and institutions, journals and books related to training research; and a project management information service.

One last facet of CMI that warrants mention before we move on is where courseware development programs are designed to evaluate the effectiveness of ICAI frames. CMI keeps track of and compiles the responses associated with a particular frame to assay if it is effective in the learning process. It assists in designing programs.

More Power Through Peripherals

Computer power in the self-learning process is enhanced by peripherals. Many of the peripherals soon become incorporated into the hardware unit; others are in the form of software, still others are up in space. Let us explore the world of computer peripherals and their implications to the areas of distance, storage, visuals, voice, and courseware development.

Each day, our world becomes smaller through the installation of new communication devices. Satellites go up and another segment of the world's population is introduced to a deluge of information. Solar batteries get married to microwave towers and information flows over the air. A modem marries a telephone receiver to a computer and in a few short minutes a vast amount of information is exchanged and stored for recall and human assimilation. If the human is not present upon the receipt of the message, the computer will hold it in the form of electronic mail. Cable and satellite transmission is redefining the concept of distance learning both in the developed and developing countries. In the U.S., the American Association of Junior Colleges has approved the formation of a community college satellite network to share teleinstructional materials.

Nova University in the state of Florida, U.S.A. is pioneering courses via electronic mail and computers with students residing hundreds of miles from the campus. Electronic mail in the form of lessons and instruction is exchanged in this process. The degrees are accredited.

The Open University of Thailand with about 225, 000 students is in the process of developing interactive computer packages as part of its strategy, in addition to satellite transmission (Oliveira, 1988).

Thirteen Indonesian Universities are tied together in a satellite network that allows them to share resources; another network unites the countries in the Pacific Rim Islands; still another ties together those island countries in the Caribbean Basin (Stoddard, 1987; Tietjen, 1987).

Since the Intel Satellite Program became a reality in 1964, it has diversified services to include 1800 channels and serves subscribers in Latin America, Africa and Asia. It is experimenting with voice radio to reach remote mountain and desert areas of Algeria. Its goal is to make global communication a reality. Intelnet, upon completion, will connect large earth stations with microterminals (Carter, 1987). Texas Instruments recently conducted a teleconference through satellite that included 50, 000 people in one hook-up (Thompson, 1986).

The satellites are up, the micros and communication ties are being installed. It is difficult to say which is peripheral when comparing satellites to computers. But whichever the posture, both are redefining the distance learning and self-instructional processes.

Rapid advances in the area of storage peripherals are taking place. Just a short time ago, a megabyte or one million bytes of memory was common. Now the common memory space is forty megabytes and the memory has been made to fit into a laptop computer (Unger, 1989). Because of the vast amounts of storage required for graphics, this is an important breakthrough in the area of courseware development. But of even more significance is that of compact disk storage, which will accommodate over 250, 000 pages of information on a disk five inches in diameter. In addition, it can recall any information in two or three seconds (De Tray, 1986).

Another technology to hit the market square soon is called DVI, for Digital Video Interactive. This 5-inch compact disk will store 3,000 still frame images or photos, up to 2 hours of quality stereo sound, or about 600 megabytes of data (Ofiesh, 1986; Angus and Waurzyniak, 1987). It is evident that cassette video recorders are on the way out and video disk cameras will appear in the market.

The competition and pace of the peripheral market is intense because of the market potential associated with this aspect of the information revolution. What it means to the training arena is that libraries can now be carried in a briefcase. In courseware development, it means the pictorial inputs can be made concrete from actual real-world situations rather than in more abstract graphic interpretations.

Another technology that is fast approaching the video scene is that of fiber optics. There is now a new electronic visual device that can "see" patterns and diagnose maladies from a color-and-space perspective using the computer as its processing and memory unit (Podmore and Fagug, 1986).

Already, voice synthesizers are in use so that the computer can talk and output information by sound. Now Bell Laboratories has developed a recognition peripheral that completes the communication loop. It can recognize and respond to a limited vocabulary when someone talks to it (Tymony, 1989).

All of these peripherals can and will be integrated into courseware development for the self-learning process. But there are still others that are more directly involved through new computer speciality boards and associated software.

A board developed by AT&T, called the TARGA board, allows designers to catch video frames and make slides (Neeley, 1988). The system does essentially the same thing to photographs that PC Paint software program does to graphics. Once the picture is caught, it can be clipped, subjects and backgrounds moved, altered and even undressed, all with an image processing software system. Captions and labels can be added at a keystroke.

Another technology in the area of graphics is the new color scanner which can lift photographs off paper or prints and they appear immediately on the screen to be stored in the TARGA photo file (Goren, 1989)

Another peripheral that will drastically alter self-learning programs is the hypercard with touch panels and windows. Hypercards allow the learner to explore information in a lateral rather than sequential context. Exploration is through icon or graphic menus. It may also involve what is termed hypermedia or a multitude of media profiles. A more apt description comes from the latest issue of the Computer-Based Training Journal (Manebretti and Mahler, 1989).

You pop an encyclopedia disc into your CD-I player and sit down in front of your TV with an ordinary remote control. You are interested in information on Japan since you are planning a vacation. After the encyclopedia title screens, the program presents menus from which you can select items by pressing a remote control button. You select the find option—this displays as alphabet from which you input Japan by selecting the letters J-A-P-A-N. Then you select Enter. The screen instantly displays an animated title page of the article on Japan and the Japanese National Anthem begins to play. Following the animation, you see a full-screen, high-resolution photographic image of a temple in Kyoto with graphic overlays of icons. Through prior experience, you know the icons signify language, culture, economy, and so on. You select the icon for language. The screen splits into four quadrants. The upper left hand square displays a motion video clip or a Kabuki dancer while the remainder of the screen shows a text menu with "currency", "the arts", "food". You select "food" and the Kabuki dancer is replaced by a cartoon-like animation of a pair of chopsticks lifting a tidbit of sushi. A voice in English says "May I have the menu, please? What tastes best?". You select the Japanese soundtrack and restart the sequence. A native Japanese speaker repeats the phrases you have just heard, but in Japanese, of course. You rewind the sequence, select "English", and then, just as the narrator says the word "Menu", you select "Japanese". The Japanese speaker instantly says "please" in Japanese — that's real-time! Now you return to the previous menu screen showing the temple in Kyoto. You select the icon for economy, and the temple is replaced by a still photo of the Tokyo Stock Exchange. And so on.

What happens with hypercard and the employment of hypermedia is that the cognitive psychologists all shake hands. Learners can now *discover* information like Bruner intended, through random investigation of Ausubel's semantic net. And when they get to a subject, it reverts to Skinner's sequential programmed instruction. Thus, ICAI with hypercard has the exciting possibility of applying the principles of learning psychology simultaneously in a multisensory mode. You talk about impact and inputting of information in an unprecedented fashion; it's upon us.

Imagine having a vast storehouse of knowledge at your fingertips and being able to explore it in a random intuitive fashion. Almost like working a crossword puzzle, you do the words wherever your interest lies. And it shows you pictures of your points of interest, or video clips. And it talks to you and you must touch it to get it to work. How much more personal can one get!

These peripherals are fertilizing the domain of teaching problem solving through computers. This is the edge of the technology. Once the knowledge ledger is in place and readily accessible, we have an information source for problem solving. One of the buzz words of the day in computer science and courseware development is Expert Systems.

Expert Systems

There are four basic steps in implementing the learning process and it culminates in solving problems for ourselves and societies. First, to input information there must be undivided attention without time sharing with other events. Research shows multisensory media facilitate that process. Next, we input information from some knowledge source. Third, we process that knowledge, imbibe it and store it in association with relevant experiences, memories of past acquisitions of knowledge. And most important, we call on that knowledge pool to solve problems.

The challenge to teachers is not to input knowledge through some antiquated lecture scenario (we know that programmed instruction does it better) but to teach problem solving. Yet, every time people want a problem solved, they go to the men of wisdom, the expert. How do we shorten the turn-around time of developing experts? Experts are in demand and the cloning time is too long. Expert systems is one method that holds potential for the upstream learning scenario.

Expert systems are computer programs that simulate the problem processes of intelligent human behavior (Denning, 1986). It consists of three parts. First, there is a knowledge generator that contains an array of facts and heuristics about a given domain; next, an inference engine containing the inference rules that are applicable to the knowledge base. These rules are usually in the form of an "IF-THEN" statement—IF this condition, THEN do this. Employing search strategies, inference rules are applied to the knowledge to provide solutions to problems. Finally, there is a component called an interface, the facility that allows the user to communicate with the expert system residing in the computer.

During consultation, an expert system behaves like a human expert in a problem-solving task. It asks questions to accumulate the necessary information. It explains its reasoning when prompted and justifies its conclusions when challenged. It even uses incomplete information by making logical assumptions from available data, much as the human expert will do in defining and solving a problem. (Levine, *et al.*, 1986).

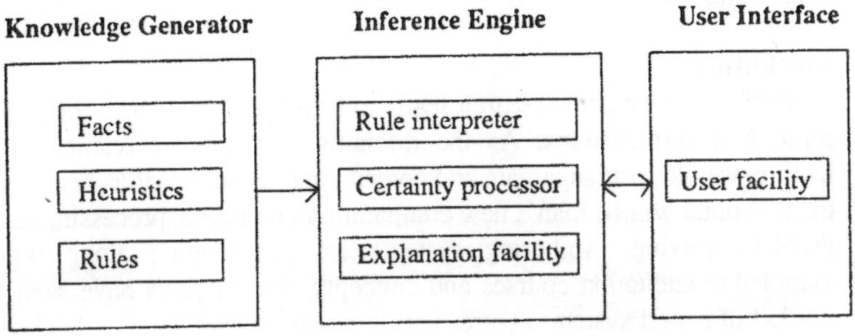

Fig. 5: Components of an Expert System

It took 2 decades—from 1965 when Dendral helped chemists identify molecular structures from spectroscopic analysis to the present, —for the expert systems discipline to gain popular acceptance. But the present growth rate of expert systems technology is drastically increasing. A study reported an increase in number of expert systems in operation from 50 to 1400 and those under development from 2500 to 8500 for the period from 1987 to 1988 (Reddy, 1989).

In response to the widespread use of expert systems, there are now more than 60 expert system development tools available in the U.S. for the PCs alone, with prices varying from $ 40 to $ 70,000. Other equally interesting tools are also available in Europe and Japan.

Many of these tools are new. Others are enhancements with new features added or old capabilities improved (Harmon, 1988).

Choice of expert system development tools ranges from simple rule-based shells or languages to versatile and powerful development environments. Rule-based shells are often restricted to a narrow and specific type of application. Some environments require the use of special languages like LISP and PROLOG. However, these languages are difficult to master and require extensive programming, which make the development effort complex and costly. The most expeditious development environments are generally shells which possess the power and flexibility of specialized languages but are easier to use (Harmon and King, 1985).

Features considered in evaluating expert system building tools are: methods of knowledge representation, type of inference mechanisms, developer interface, user interface, and external interface. Combinations of these features provide a wide range of expert system design considerations.

The Future

We know upstream research leads to artificial intelligence, perhaps through expert systems. As the domains of computer science and neuropsychology amalgamate and contribute to educational technology, the computer will be man's new companion in inputting, processing and problem solving. Authoring systems are proliferating along with knowledge elicitation courses and concepts. Recent years have seen a myriad of expert systems centres springing up all over the world. There is now an Association of Artificial Intelligence and a new Journal of Knowledge Engineering (Reddy, 1988).

The competition is keen and the stakes are high as institutions and governments embark on this new frontier. Some excerpts:

> The Japanese Ministries of Education and International Trade have embarked on a project involving 60 companies to develop courseware in Educational Computer Technology. One primary aim is to develop a cost-effective satellite dish. Another is to standardize software and make it user-friendly. Their research has found that four- to seven-year-olds can learn to operate the computer as an information device. Their conclusion is that this new teacher allows them to be able to learn by themselves if given the appropriate software (Anderson, 1985).
>
> Another whole new field is evolving in computers called Neural Network Systems [NNS]. Computer scientists and electronic engineers are mimicking the neural networks of biological organisms. One field in this area called Electronics is fashioning artificial retinas and cochleas and other biological sensing devices out of silicon. Using special analogue amplifiers, resistors and transistors and the physics underlying biological signal processing, the scientists have succeeded in reproducing the

functions of living retinas of lower animals and have produced a neural network that can teach itself to talk (Bien, 1988).

- Again Japan, as it's Fifth Generation Project in computers, is developing a computer that will allow immediate translations during telephone conversations (Kobayashi, 1986). In essence, this would contribute to a language-neutral world.
- IBM is working on a method to input handwriting into a computer from a digital pad, eliminating the keystroke as a communication link with the computer (Blaine, 1986).

  The future will see automobiles driven by remote control on electronic highways and traffic flow will be shown on a digital display screen in the car (Brand, 1987).
- The Hughes Corporation is developing a palm-sized supercomputer that has processing speeds in excess of 100 billion operations per second. It stacks integrated circuits and uses tiny microbridges for interconnectivity. It is called the 3-D Computer (CALS Technical Bulletin, 1989).

Fig. 6

On and on goes the microrevolution as it gains momentum and impacts our lives. All these projects will have profound effects on the availability of knowledge and the way people will learn in the future. But one might ask, what about the revolution and its impact and applicability in developing countries' learning systems?

India has a project to advance computer literacy throughout the country. The Class Project, as it is called, is implemented in 250 schools. It teaches computer awareness, applications, and most of all encourages teacher usage. It has created 42 resource centres to promote, maintain, develop curricula, train and assist teachers in the technology. A bachelor of computer science is available in 22 colleges of the country and the computer society of India has over 6000 members.

The Philippines, Indonesia, and Thailand all have concentrated training of trainers programs in the use of computers in the learning process (APEID, 1985). Some Vietnam Colleges of Agriculture have translated software packages and are already using the computer as a training device. Singapore University has a Think Tank on expert systems and the government has dedicated resources to upstream research and development (NCB, 1988).

As volume of production goes up and costs come down the computer and its assorted peripherals will work for everyone in the so-called developing and developed countries alike, much as the radio and television technologies that have gone before. They are already altering education and training strategies around the world.

A Parting Scenario

Why not a prefabricated information center for villages, remote or otherwise, that could serve the needs of the community, be it extension services or schools? The modular building could be transported on wheels to its site of dedication. As the community and needs grow, new modules could be added. Each module would be self-contained and engineered to produce and replicate both verbal and pictorial information much as has been done with the art of fast-food restaurants in many areas of the world. The energy could be supplied by solar panels. But of course, it would be fitted with a satellite dish to receive information from the outside world (Energy Technology Update, No. 013T).

New time sharing of satellite windows makes subscription cost-efficient.

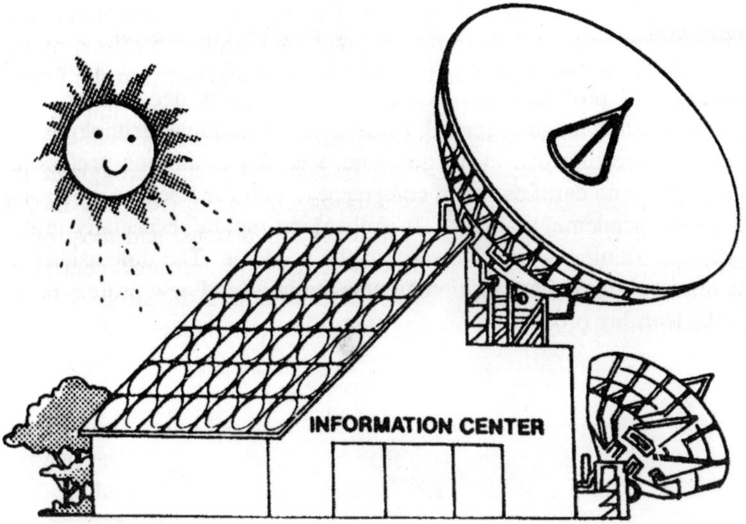

Fig: 7

At a certain time of day, the information drop could be picked up and stored in the computer, printed onto paper, posted at the market place, distributed to the school and taken home to the parents by the school children, much like a daily newspaper. The information drop could contain bulletins about weather predictions, seed and fertilizer availability, the extension lesson for the week in the health and agriculture areas, current events and simple self-learning information packages.

Extension services and the school systems are both in the information and learning business. However, in many parts of the world, what is lacking is a support entity to disseminate information. Information is imperative to decision-making.

Distance learning is coming of age with the microrevolution. The pieces of the support entity are already available in many parts of the world. Perhaps it is time to introduce the dimensions of the microrevolution to the learning process in both urban and less accessible areas.

This microrevolution has no preference for culture or discipline domains. It sprouts in Bhutan as well as Britain. It works in India as well as Indiana. It is used in medicine as well in business, and the dimensions of this revolution are starting to impact the learning process. It is changing learning environments and creating new domains of learning and professionals unheard of a few years ago—courseware designers, knowledge engineers, visualizers, educational technologists.

It is interesting that in the academe, teaching is the one profession that requires no certification of competency. You get a degree, you get a job in the academe, you teach. It is changing rapidly, especially in the corporate training world and soon in the academe. The dimensions of the microrevolution are fashioning new methods and new professionals for the learning process.

References

Anderson R. H. (1985). Expert systems: Practical applications and development. The Artificial Intelligence Symposium 1985. Marco Polo Hotel, Singapore.

Angus, J. Waurzyniak P. (1987). RCA interactive technology puts full motion video on CD ROM. Infoworld News Vol. 5.

APEID, (1985). Computers in Education: An Outline of Country Experiences. UNESCO Regional Office for Education in Asia and the Pacific, Bangkok, Thailand.

Bien, B. (1988). Promise of neural networks. Am Scientist. 76(6): 561-564.

Blaine, D. L. (1986). IBM's Writing to read program. Paper presented at the World Conference on Technology and Education. July 28-August 1, 1986. Philippine Plaza Hotel, Manila, Philippines.

Brand, S. (1987). The Media Lab. Viking Penguin Inc., New York, NY.

CALS. (1989). Continuing trends in miniaturization. Centre for Advanced Learning Systems Technical Bulletin No. 2:1.

Carter, W. G. (1987). What Intelsat could do for development. Media Development. 34(1): 38-39.

Community College Satellite Network is Established by AACJC (1989) Telectrends. 4(3):4.

De Bono, E. (1970). Lateral Thinking. Penguin Books, New York, NY.

De Tray, J. (1986). CD-ROM: A Curiosity No Longer. CD-ROM Rev. 4.

Deevey, E. S. Jr. (1960). The human population. Scientific Am. 203(3): 194-204.

Denning, P. J. (1986). The Science of Computing: Expert systems. Am Scientist. 74 (1): 18-20.

Extension Committee on Organization and Policy, (1982). The Computer Management Power for Modern Agriculture. Cooperative Extension Series. Purdue University. Lafayette, Ind.

Fisher, A. E. (1964). Chemical stimulation of the brain. Scientific Am. 210 (6): 60-68.

Goren, B. N. (1989). Professional products. True Imaging. 2 (1): 15-20.

Harmon P. King D. (1985). Expert Systems; Artificial Intelligence in Business. John Willey & Sons, Inc., New York, NY.

Harmon, P. (1988). Currently available expert system-building tools. Expert Systems Strategies. 4(7): 3-14.

Hermann, N. (1988). Brain Books. Lakeshore, North Carolina.

Hewlett-Packard, Southeast Asian Division (1986). School and Institution Discount Program.

IBM, Philippines. (1985). Hardware Grants Program to Universities and Institutions. Manila.

Kobayashi, K. (1986). Computers and Communications. Massachusetts Institute of Technology, Cambridge, Mass.

Knowles, M. (1984). The Adult Learner: A Neglected Species. 3rd ed. Gulf Publishing Co., Houston, TX.

Levine, R. I. Drang, D. E, Edelson, B. (1986). A Comprehensive Guide to Expert Systems. McGraw Hill, Inc., New York NY.

Mager, R. E. (1962). Preparing Instructional Objectives. Fearon Pub. Co., Belmont, California.

Manebretti, C, Mahler, P. (1989). CDI: A Preview. CBT Directions. 2(6):19-20.

Minnick, D. R. (1984). Microcomputer applications in agricultural research. A workshop at the International Rice Research Institute. Sept 24-27.

Minnick D R (1988) Computers in the learning process: their use in human resource development and management. Paper presented at the International Conference on Research in Plant Science and Its Relevance to the Future. University of Delhi, India.

Minnick, D. R. (1989a). A Guide a Creating Self Learning Materials. International Rice Research Institute, Los Banos, Laguna, Philippines.

Minnick, D. R. (1989b). The knowledge generator. NACTA J. In press.

NCB, (1988). Singapore IT'87 survey. National Computer Board, Singapore.

Neeley, D. (1988). Architecture and image capture. True Imaging. 1:33-36.

Novak, J. D. (1977). A Theory of Education. Cornell University Press, Ithaca, NY.

Ofiesh, G. D. (1986). The Seamless Carpet of Knowledge and Learning. Pages 299-318 in CD ROM: The New Papyrus. Lambert and Ropiequet, eds. Microsoft Press, Redmond, WA.

Oliveira, J. (1988). Trends in distance learning; A new wave. Development Communication Report. 63:1-4.

Owen, S. V, *et al.*, (1981). Educational Psychology: An Introduction. Little, Brown and Co., Canada.

Podmore, C, Fagug, D. (1986). The challenge of optical fibers. Telecommunications Policy. 10(4):341-351.

Pribram's paradoxes: How does the brain know? (1977). Brain Mind. vol 1 & 2. p. 2.

Reddy, R. (1988). Foundations and Grand Challenges of Artificial Intelligence. AI Magazine. 9(4):9-21.

Stoddard, R. (1987). Developing the third world. Satellite Communication. Jan.: 20-21.

Thompson, P. (1986). Teaching through video conferencing. Communication International. 13(11):43-45.

Tietjen, K. (1987). A.I.D. rural satellite program: An overview. AID Rural Satellite, Washington, D. C.

Toffler, A. (1980). The Third Wave. Morrow & Co., Inc., New York, NY.

Transportable solar pack for remote aboriginal communities. Energy Technology Update. No. 013T, p. 1.

Tymony, C. (1989). Is your computer listening/CBT Directions. 2(5):18-20.

Unger, J. (1989). A pair of sophisticated laptops. Byte. 14(2): 189-194.

Wales, C. E, Stager, F. A. (1977). Guided Design. Center for Guided Designs. West Virginia University, Morgantown, West Virginia.

Whittaker, A. D, Thieme, R. H. (1989). Integration of problem solving techniques in agriculture. AI Magazine. 10(2):85-87.

7

Issues in Research Utilisation and Programme Evaluation in Development Communication

Ronald E. Ostman

Many journalists follow a simple formula when they write their stories and articles. It is the "5W and H". Simply, a good journalist tries to answer the *Who, What, When, Where, Why* and *How*. The journalist also considers which of these six elements is most important. Because the reader, viewer, or listener often is busy, inattentive, and impatient, the journalist tries to state the most important of the elements first and to reemphasize those elements in subtle ways in order to reinforce their importance. As I was trained as a journalist before I was trained as a researcher-evaluator, I have always found it useful to apply the 5Ws and the H in research and evaluation as well as in journalism. Therefore, this exposition will be organized according to the 5Ws and H. The following words will not be an exhaustive catalogue of issues facing evaluators in development programmes today. Rather, I have selected those issues which I have observed most frequently and which persons working in development programmes have often raised in discussions.

But, before beginning, let's pause to examine what will be meant by evaluation research. We do not live in a static world. Change occurs whether or not we want it. In fact, often the change in unintended and unwanted. But it occurs anyway. In response to the inevitability of change, some of us sometimes try to direct change (Nunnally, 1983). We seek to control ourselves, our world, and our fellow humans. Often, these attempts to control are half-hearted, half-cocked, haphazard, and half-witted. Rarely, our attempts are well thought out, wise, amply supported by adequate resources, etc., giving them at least some chance of meeting expectations and goals. But, almost never are these rare attempts adequately evaluated. That is, do we use good information to plan social change? Do we have an accurate picture of existing reality? Do we use data to formulate our strategies and change methods? Do we test our strategies and methods before committing ourselves to them? In communication, do we pilot test our messages and media before delivering finished products to audiences to whom we wish to communicate? When we implement our projects and programmes, do we monitor them for effectiveness and whether or not they go according to plan? When our projects and programmes are completed, do we know the extent to which they succeeded in meeting objectives? Do we communicate these successes, problems, etc. to appropriate persons? Are our data used to assist in project and programme revisions in the next cycle of change attempts? If the answers to any or all of these questions is "No", then we have not been evaluating.

Evaluation. Those of us who are attempting to control our environments, our own activities to improve our situations and quality of life (and situations/quality of life of those with whom we share contemporary time and space) should evaluate, because if we do not, we are practically guaranteeing failure. Not only that, we are ensuring that our failures will be repeated in the future. Evaluation information will help tell us what we should be doing, how we are doing, and what we have done wrong in the past. Definitions of evaluation imply this. The primary authors behind what probably has been the most successful evaluation guide (or "kit" as they refer to it) in 20th century publishing history, Marris *et al.*, (1978a) and Herman *et al.*, (1987) define a programme as "anything you try because you think it will have an effect" (p. 8) and evaluation of a programme when "you systematically collect information about how the programme operates, about the effects it may be having and /or to answer other questions of interest" (p. 8). This information ("data") should be credible, valid, reliable, useful and should make a difference.

Who should Conduct Evaluation Research?

Traditionally, "experts" conducted evaluation research. These experts were well educated, particularly in the wiles of social science. They usually were trained in the West or uncritically adopted methods developed in the West (Campbell, Shrestha, and Stone, 1979). Often, even though working in developing countries, they were Westerners. Almost always they were credentialled with Ph.Ds, Ed.Ds, or at least master's degrees. Nearly always, they represented the authorities, such as the government and its bureaucracies. Often, they were "outsiders" to a project or programme, sent in to make sure that resources (money, persons, materials, equipment, etc.) were being used in the ways in which they were intended and were not being wasted. Usually, they did not stay around long enough to be anything but strangers.

That these persons often were not capable of conducting *bona fide* evaluations became painfully obvious. They often were superficial, ignorant of context and situation, nonconversant in language and custom, and generally "babes in the woods". It was easy for project/programme officials to dupe them, or at least to steer them into areas where they would do no harm. They often focused on the wrong things, talked to the wrong people, saw the extraneous, and in general got in the way. They probably "came at" the task of evaluation in ways which fundamentally differed from the cultural predilections within which they found themselves. For example, Dissanayake (1987) has speculated that the very basis of thinking about communication varies between Western and Indian cultures, saying that Westerners conceive of communication from an expressive point of view, whereas Indians are more comfortable with interpretation or reception.

However, the outsiders had at least one major advantage over the "insider" evaluator, the person who was intimately associated with particular projects/programmes because he/she was a routine worker with the social action plan. That advantage was the potential for at least a modicum of objectivity. The insider evaluator tends to become a member of the team and supportive of the project/programme. In most kinds of evaluation, this is a positive characteristic. For example, the supportive evaluator will collect needed information and will share it with the appropriate peers and colleagues in order to improve the project/programme in operation. But the insider evaluator, who has friends and a job to protect, does not appear to be well suited to be critical of a project/programme which has floundered badly, or even which only has wobbled slightly. This, however, potentially can be done by a relative stranger, an outsider who may be able to spot flaws that insiders have learned to live with or chose not to see. Particularly in summative evaluation, when the major information quest is to find out whether or not various project/programme objectives and goals have been reached *after* the project/programme largely has been concluded, an outside evaluator is desirable. Often, whether or not a project/programme has been at least partially successful will determine whether or not it will continue to receive personnel, funding, equipment and other resources. It would take a very open-minded, modest, probably masochistic insider evaluator to arrive at findings which pointed toward the logical demise of the project/programme.

But, whether we are talking about insider or outsider evaluators, we still are considering evaluation from a "top down" perspective. That is, relatively high status, elite individuals are administering a project/programme for the presumed benefit of "clients" (for example, villagers, rural people, marginal farmers, Harijans, or whomever). What is considered desirable, necessary, and beneficial by those on the higher rungs of the hierarchical ladder provides the substance of the project/programme, the way it is run, and so on. In top down projects/programmes, the client is rarely, if ever, consulted, either as a partner in the project/programme, or as anything but a passive source of information and person to be manipulated. The need for informational substance is decided upon by nonclients (sometimes called "providers").

Concurrent with a development perspective which insists on "bottom up" strategies and approaches, some evaluation researchers have begun to consider whether or not the clients ought to share in the information collection and consumption process. Those researchers who have chosen to include clients as colleagues subscribe to a concept of "participatory research". For example, in adult education, participatory research originated in Tanzanian villages in the early 1970s "based on the principle of the villagers themselves being active participants in a research plan which would ultimately motivate them to evaluate their own strengths and needs for development of their own villages" (Yarmol-Franko, 1988, p. 3). Participatory research consciously departs from dominant, establishment science. It rejects neutrality and objectivity as myths. Participatory research celebrates subjectivity. It seeks to oust elitists from research control. It glorifies "ordinary people" (the deprived, oppressed and under-privileged) and seeks to place them in control of not only the research process and its resultant information earned through knowledge production, but over decision-making and implementation of projects/programmes which will transform their daily lives. Thus, the end goal of evaluation for the participatory researcher is not the time-honoured "search for truth" via dispassionate detachment (Cohen, 1982), but answers to questions of daily survival and reduction in the struggles of ordinary people. Information has value largely because it is a source of power and potential action. Gone is the belief that the mental and the manual should be separate; in fact, the intention is to fuse them. Also, the participatory researcher seeks to heighten emotion, not to render feelings immobile or in obeisance as the establishment researcher might wish. In a sense, then, participatory research is but a means to an end. By "interfering" only in the sense of assisting ordinary people with certain technical skills and requested advice, the participatory researcher turns over control. Ordinary people make the decisions about research inquiries, about collecting data, about analyzing data, and about what should be done in response to outcomes of the research process. The rhetoric of participatory research celebrates many cultural traditions, seeks to bolster ordinary peoples' self-confidence, tries to make knowledge accessible by wrenching it away from the greedy claws of the dominant elitist system, fosters resistance to the forces of domination, and glorifies the liberation of peoples' minds. In the end, it promotes collective processes, rather than individualistic pursuits (Tandon, 1988).

Even if evaluators do not choose to directly involve ordinary people in their research process, they can share findings with these individuals, who can be of enormous help to the evaluator. For example, sharing puzzling findings with those from whom data were gathered may permit them to share with evaluators personal insights as to why things turned out that way. In other words, they can provide meanings for data which evaluators often are not capable of producing. This boils down to a realization that while evaluators often are sensationally adept at producing information, they oftentimes are weak to explain just what the data mean. Their sense of understanding about the data is too often weak. A reading of evaluation reports often leads the reader feeling inundated with numbers, statistics and facts, mystified as to what was learned.

What should be Evaluated?

Many, many evaluation studies have been founded on the theoretical assumptions of KAP—that certain kinds of Knowledge leads to certain kinds of *A*ttitudes, which in turn lead to certain kinds of *P*ractices. From extension workers promoting agricultural innovations to public health workers trying to immunize children against disease, the project/programme often has assumed that once clients come to know something, attitudes and behaviours will fall in line; that is, will become consistent with knowledge. A result has been stimulus-response planning, which emphasizes the production and dissemination of planned stimuli, which almost certainly will lead to desirable effects. Or, will they? Common sense suggests that not everyone pays the same amount of attention to stimuli, even assuming that stimuli are equally available to everyone. And even if everyone had equal access and paid equal attention, their interpretations and perceptions of the stimuli would be likely to vary (Britt, 1978). Moreover, only modest relationships have been found in social scientific studies which have sought to link, for example, attitudes to behaviour (McPhee and Cushman, 1980).

So, other than KAP, what is there to study? Other factors deserve our consideration as we gear up to study evaluation. Britt (1980), for example, points out the usefulness of such general areas of psychological human nature as motivation, perception, lifestyle, personality, memory, anxiety, intelligence, etc. Others, with a more social psychological or sociological disposition, point out that the primary and secondary groups with which persons associate have powerful influences on individuals (Babad, Birnbaum, and Benne, 1983). Representative groups include sex/gender, family, friends, occupational groups, racial groups, religious affiliations, political contacts, age life stages, etc. Still other possibilities for evaluation studies include needs and wants. Self-concept and self-esteem. Values and beliefs. Public opinion. Communication networks. Organizational aspects of human existence. Ideologies. Status. Social class. Power. Goals. Decision-making. Conflict and cooperation. Genetic/hereditary characteristics. Social situations. Nonverbal communication. Formal and informal message production and consumption. Work routines. Hobbies and leisure time use. Gender. Ethnicity and race. Education. Cultural influences. Programme operation and performance. Impact of messages. We could continue this list almost indefinitely, since it reflects the concepts and concerns of the social sciences in general—political science, sociology, social psychology, psychology, anthropology, marketing and economics, education, and so on. Some of these concepts, obviously not all, should and can be examined in the course of a given evaluation. The evaluator must choose from among these and other concepts, particularly if he/she is working deductively, early in the research process. Even inductive and qualitative research demands some sense of what is of most interest and what is worthy of most attention. Knowledge, attitudes and behaviours are important, of course. But they are not the only things which are important to our projects and programmes. Let us be creative and imaginative in deciding what we shall study.

When should Evaluation Occur?

Traditionally, only summative evaluation has been done. After a project or programme has been concluded, someone asks whether or not it had any effect and an effort is made *ex post facto* to determine the outcomes or consequences. Sometimes, such an evaluation has real value. For example, there may be plans to extend a project or programme, and some assessment is needed to decide whether or not it should continue, and if it should, what has been accomplished and what still requires work. But, the disadvantage of a summative-only evaluation approach is that the information it yields is available too late to modify any mistakes and failures that occurred in the past. Evaluation should begin at the beginning! In a goals/objectives-based project or programme, for example evaluators should assist in setting those goals and objectives. If a project or programme has no goals or objectives, or if they are implicit, assumed, or poorly articulated, it will be very difficult, if not impossible, to determine the effectiveness of the project or programme (but from a different perspective on this, see the "goal-free evaluation" approach of Scriven, 1972). The goals and objectives must be stated in a form which permits the evaluator to know what the project or programme hoped to accomplish, in order that measurements can be made of actual progress toward meeting those goals and objectives. The evaluator is the best possible position to know in what form the goals and objectives need to be stated (Morris and Fitz-Gibbon, 1978b).

Evaluators also can assist project/programme planners by conducting needs assessments (Neuber *et al.*, 1980), audience profiles (Doshi and Agrawal, 1989), and situation analyses. Often, little information is available concerning specific individuals and communities. What do they want and need? What problems do they face? What are their levels of knowledge? Their feelings? Their behaviours and achievements (Morris and Fitz-Gibbon, 1978d)? Their motivations? Their perceptions? It is helpful to have answers to these questions and others *before* designing projects or programmes. Evaluators also should work with formative evaluation. This means that they will help project/ programme staff; for example, they may help develop and pilot test messages (Haaland, 1984). They will monitor projects or programmes in operation, making sure that things happen when and as they are supposed to happen (King *et al.*, 1987; Morris and Fitz-Gibbon, 1978c). They will keep surveillance on the social and natural environment, watching warily for any interferences with the project or programme when it is in operation. In short, long before summative evaluation occurs, evaluators can have been of immeasurable help in making sure that a project or programme is well conceived, well planned, well executed and well received. Evaluators can thus operate as individuals who bring helpful and good news, in time to be of use in making adjustments, rather than just being potential threats who bear bad news, too late to do anything about it.

Another time issue in evaluation research (in fact, in all social science research has to do with when data ought to be collected. For many years, social scientists and evaluators have been content to take "snapshot" pictures of humans, thus giving a static portrait. But humans and life are dynamic. If we are interested in change, we need to observe change occurring over time. We need time comparisons. For example, what were attitudes and knowledge before the project or programme? What were attitudes and knowledge during the project or programme? And what were they after the project/programme? To answer these questions, and to assess whether attitudes and knowledge changed, we'd need at least two measures of relevant attitudes and knowledge levels. The time element is neatly incorporated into many "pre-post" experimental designs (Campbell and Stanley, 1963). However, because other methods also are used to collect information during evaluations, we need to be conscious of the time element with all our information collection procedures, such as content analysis (Holsti, 1969), survey research (Warwick and Lininger, 1975), participant observation (Agrawal, 1975), etc. The idea is to create a moving picture, or at least a series of comparable still photographs, rather than just rely on a single instance in time.

Within a longitudinal series of researches, a further issue is whether the evaluator ought to ask for/take information from the same individuals at different points in time (a "panel") or whether different samples of persons "trend studies" or "cohort studies") ought to be examined over time (Babbie, 1986). If one uses a panel, it is justifiable to talk about change or lack of it, because the evaluator is comparing a person to himself or herself as time elapses. If different samples are used at different points in time, it is still possible to talk about change, but the evaluator has to make assumptions (for example, that the groups were equivalent at time one and that any changes observed in the second group at time two also occurred for group one, which did not share data at time two). One problem with a panel design is the possibility that the persons who shared data also were "sensitized" as a result of having given that data. For example, if we ask a farmer about his or her knowledge concerning some new agricultural technologies or techniques, he or she may realize that his or her knowledge is poor or incomplete on the technologies or techniques of inquiry. Given a subsequent opportunity, our farmer may be sensitized to learn the new information. In other words, he or she may learn about the new machines or procedures as a function of having been made aware of his or her ignorance at the time of data sharing and collection. Thus, a subsequent measurement of knowledge on these topics would show gains, but the improvements would be due, at least in part, to the fact that we asked about them at time one. Knowledge gain would be a function of having participated in the data sharing/collection, whereas we may be inclined to attribute gains to our projects or programmes. At the same time, those farmers who were not sensitized by the data sharing/collection process may have ignored the information made available by our project/programme. Thus, we'd be in error if we attributed the knowledge gain solely to our project/programme. Of course, evaluators have worked out ways to reduce the possibilities of sensitization in panel research designs (for example, by disguising their inquiries, or by asking for alternative data which are thought to be comparable), but the danger of sensitization persists. More elaborate research designs, such as the Solomon four-group design, permit evaluators to assess the probable impact of pretesting and sensitization (Solomon, 1949; Triandis, 1971). Other problems with panel designs, such as people dropping out (for example, due to lack of cooperation, shifting residences and locations, death), and changes in the meaning of research instrument items over time have been mentioned by Hyman (1972), Kish (1965), and Markus (1979).

Where should Evaluation Take Place?

In historical research, students are urged to study "primary" documents and evidence rather than "secondary" resources (Smith, 1989). A primary historical document is an eyewitness account of the past, or it is an actual artifact of the past. A secondary resource, on the other hand, is illustrated by someone's record of what an eyewitness saw, recalled and written by the former, say, some 50 years after the actual event. The principle which we can extract from the historigraphical advicé is this: always deal directly with the phenomenon of interest if at all possible. An observer who is interested in a village's agricultural practices should go to the village, seek out farmers, and visit their animals, their fields, their aviaries. If words are to be collected as valuable bits of information, the farmers' own words should be sought and preserved. Evaluation should take place "where the action is", not on the basis of a third parties' interpretation or analysis or recollections. Talk *to* farmers, not *about* farmers. Of course, it is not always possible to "get it from the horse's mouth", but when the evaluator is forced to rely on secondary sources or references, the information that is gleaned should be interpreted with a healthy dose of skepticism and caution.

The implication here is that we must be able to speak to "client's" language, both literally and figuratively. People are not floating in space. Their contexts and situations must be understood. We must, as evaluators, see things through the eyes of the client before we can make sense of our project/programme successes and failures. There are no short cuts. Evaluation, if it is to be worthwhile, must have sufficient resources and the evaluators ample motivation and stamina to go into the field, to be part of the project or programme in action. Reading books in the library about a people or their region, while valuable, is no substitute for direct and sustained interaction with them in their regions.

We are speaking of data collection. Of course, it often will be proper to plan data collection or to analyze collected data in places other than with the clients in their villages or regions. The need, for example, for dependable electricity or computers, often mitigates against staying in the field. Even in this extreme instance, however, it should be noted that evaluators are working on improved technology which permits rapid surveys. It is now possible to develop a survey design, collect the needed information, and have a written report in a matter of only a few weeks, even under extremely difficult field conditions (Frerichs and Tar, 1989).

Why is Evaluation Necessary?

The need for evaluation already has been argued. To reemphasize this point, consider an example. The Kheda Communications Project of the Development and Educational Communication Unit of the Indian Space Research Organisation began operations in 1975. Television programmes were broadcast to rural villages where TV had never before been seen. Among subjects broadcast were programmes on agriculture, animal husbandry, health, "womens' issues" (such as cooperative, economic activities), etc.

Anyone observing the community viewing of television programmes in the rural areas would have noticed many more men than women in the audience. In fact, evaluation research showed approximately a 2:1 ratio of men to women viewers. Suppose no formative research had occurred during the 10-year period until 1985. Suppose programme producers merely continued to create their messages and send them via the broadcast airwaves at times chosen rather arbitrarily in urban planning centres, considering their job to be done. How effective would the programmes have been? Several problems are immediately apparent. Messages primarily aimed at women would reach but a few women. The predominantly male audience may have become disenchanted with TV fare which stressed themes and subjects prepared primarily for women.

If the television producers were made aware of the audience composition, on the other hand, and if they were informed *why* the audience was largely male, then they could adjust programmes and the timing of programmes in order to maximize the possibility of reaching females. Further, they could take steps to ensure that more females be brought into the audience.

Evaluation research yielded just such information. Evaluators determined, for example, that workloads prevented both men and women from viewing TV. However, workload was a more important reason for men than for women. It was social taboo which restricted women's TV viewing. In rural communities, social restrictions prevented women from leaving the home during evening hours, especially to go into community situations to view programmes on a single community TV set.

Feedback studies, conducted immediately after the television broadcasts, further revealed that women preferred programming which came in a continuous play format and which stressed realistic themes on relevant issues. They liked positive endings to the stories. The best time to show such programmes, evaluators were able to determine, was between 8 and 9 p.m. Programmes which also contained messages for men, even though perhaps primarily focused on women's issues, were preferred.

Television personnel, who were largely accustomed to the urban situation and individual ownership of TV (and thus probably had no idea of the social taboos operating to limit their female audience), were able to respond to the information provided by evaluators:

> "...when women's programmes were transmitted on a regular basis at a definite time and when these programmes took into account the needs, issues, problems, and aspirations of women, the proportion of women viewers increased to about 40 per cent of the adults for those programmes [compared to an earlier average of approximately 33 per cent]. It may also be noted that while participation of women increased, there was no decrease in the number of men viewers even for those programmes" (Joshi and Joshi, 1986, p. 25).

A summative evaluation, taken after all the programmes had been broadcast at inappropriate times, would have provided the needed information too late. If no evaluation had been conducted at all, the television personnel probably would have smugly assumed that they'd done their job, when in reality they'd reached very few of their intended viewers. Evaluation is useful, because it provides alternatives and information for decisions (Edwards, Guttentag, and Snapper, 1983). This is particularly so when evaluation addresses worthwhile questions in a context which enlightens *all* participants in a democratic, pluralistic process (Cronbach, *et al.*, 1980).

"Why evaluate"? This question can be approached from a slightly different viewpoint than previously. That is, should the evaluator evaluate as a member of the establishment and status quo? Or should the evaluator be a "boat rocker", a person who stirs things up, who challenges the existing order, who looks for problems which most others would just as soon let rest? Evaluation is almost always done to justify and to shore up the existing way of doing things, in support of the existing paradigms. Evaluators, operating within this framework, are expected to help buttress what is. If change necessary, the usual approach is "Let's change it just enough so it gets back on an even keel—no more, no less". This is evolutionary change at best. But sometimes we need revolutionaries. Sometimes the existing way of doing things, the existing social order, etc. is dysfunctional, or corrupt, or morally and ethically bankrupt. There is no reason why an evaluator cannot be part of finding alternatives, of planning, implementing and perfecting new, innovative projects, programmes, indeed, societies.

This will take courage and will cause controversy. But I have no doubt that evaluators are not in a strategic position, playing a vital role, if they choose to follow a less conservative bent. There is a saying which should be kept in mind: "If you're not part of the solution, you're part of the problem". Another saying suggests: "To take no decision and no action is to decide and to act". At the very least, we evaluators should spend some time thinking critically about the policies and priorities of the projects and programmes with which we work.

How should Evaluation Research be Conducted?

A wide range of methods, or tools for collecting information ("data") are available to the evaluator, ranging from the tightly controlled, laboratory-type experiment (Houston, 1972) to naturalistic participant observation (Agrawal, 1985; Patton, 1987). Other methods which may be considered include field experiments (Stanley, 1972), survey research (Warwick and Lininger, 1975), content analysis (Holsti, 1969), physiological measurement (Riccillo, 1989), interaction analysis (Emmert, 1989), nonverbal measurement (Donaghy, 1989), historical, legal, projective, clinical, focus groups, key informants (Ostman and Doshi, 1989; Ostman, Iyengar and Kothari, 1989), etc. Perhaps the best advice which can be given is that on one method is always preferred (Cronbach, *et al.*, 1980). The reason for this statement is that the evaluator must be free to design methods which can be depended upon to collect reliable and valid information depending upon particular situations, with particular people, at particular times. No single social science method can guarantee such information. There is no social scientific equivalent to the haberdasher's claim that "one size fits all"!

Another "how" question which probably is on most evaluators' minds has to do with resources with which to conduct evaluation. We need personnel, time, space, equipment, and (yes!) money to do a good job in evaluation. Traditionally, evaluation has been an afterthought and thus has been chronically shorted in what we loosely characterize as the "wherewithall." As a result, many evaluators spend much of their time in pursuit of resources, which naturally interferes with and reduces their effectiveness in conducting evaluations. We need to do a better job in securing sufficient resources. I have no magic solutions, but suggest that evaluators must try to have evaluation included as a "line item" in initial budgets. This will involve lobbying efforts with project and programme directors, as well as with executives in sponsoring agencies. Evaluators should be realistic in pressing their needs. Demands should not be excessive. Evaluators should ask for enough to do a respectable job, but should not "feather the nest".

An issue which we probably will hear more of involves evaluation research designs. Traditionally, evaluators have been influenced by rigid experimental methods. We have been taught that laboratory and field experiments are model research designs. Our approaches have been deductive (beginning with theories, devising methods to collect information to test those theories, and analyzing data in order to verify or modify theories). This model has worked reasonably well, but increasingly we hear from evaluators who are interested in a more inductive approach to their work. That is, they choose to be very flexible and relatively unstructured in their daily observations of projects and programmes, with the idea that in so doing they'll increase their understanding and eventually be able to build a "grounded theory". The general model for this approach has its roots in the qualitative research tradition which has been advanced with increasing vigor since the early 1970s (Anderson, 1987; Christians and Carey, 1989; Fortner and Christians, 1989; Smith, 1988).

I think the inductive approach has value in situations where projects, programmes and personnel are operating without previous experience or without a lot of prior understanding of problems, people, customs, procedures, etc. However, the dangers of a totally inductive approach are many. Without focusing, the evaluator may become lost in a thicket of phenomena. Even if he/she does focus within the thicket, there is no guarantee that what is being concentrated upon will be important. Keeping track of information gleaned will be problematic. Personal emotional involvement will likely be great, which sometimes can cloud judgment and analytical abilities.

Of relevance to this deductive/inductive consideration is the general issue of whether evaluators ought to be abstract, general and theoretical in their orientation, or whether they ought to be concrete, specific and applied in what they're doing. It is quite clear that evaluators are expected by most parties to be the latter. It's their job to describe, predict, understand and control the real world. But, I agree with others, who've noted that there's nothing so practical as a good theory (Hoover, 1988; Reynolds, 1971). A theoretical understanding does not preclude effectiveness in "mundane reality" (Anderson, 1987). Quite the opposite. A good theory makes the evaluator more—not less—effective. Therefore, I believe the evaluator can serve both research goals.

M.C. Sekhar and I have argued that a judicious blend of both deductive and inductive approaches, of both quantitative and qualitative methods, of theory grounded in reality, will yield the best results (Ostman and Sekhar, 1989). I still believe that to be the best advice. However, it is clear that sometimes an evaluator will be more qualitative than quantitative, or more deductive than inductive. Sometimes, evaluators will be confident in stating and testing hypotheses, while at other times, the research evaluation will be exploratory and tentative. The proper approach to be taken will depend upon careful judgments concerning specific situations, people and circumstances.

Related to this issue is whether or not the evaluator should be covert or overt. A covert evaluator gathers data unbeknownst to the people from whom the data are being collected. Their permission has not been sought. They suspect nothing. Under these conditions, the evaluator may feel he/she is finding out "what really happens", because people are not responding to the presence of the evaluator as an evaluator. The "demand characteristics" of the research and resultant contaminating responses are nonexistent (Smith, 1988). This, if true, means that the external validity or naturalness of what is being found is very high. On the other hand, there may be ethical and moral issues involved in such "undercover" operations. For that reason, some evaluators insist that all data collection be overt, made with the full knowledge and consent of those from whom information is sought. While admirable in terms of honesty and forthrightness, this latter approach may taint the data. People may be inclined to give the evaluator what they perceive he/she wants, to confirm the hypotheses which they perceive the evaluator is seeking to confirm or disconfirm. Generally, I believe evaluators should operate in the open, unless there are very compelling reasons to do otherwise. If careful procedures are followed, the loss to external validity usually will be small.

Another "how" issue in evaluation is concerned with *unit of analysis*. This issue has to do with the perspective which the evaluator adopts. Will the data collected be intraindividual (psychological)? Interpersonal (small group, such as family, peer, work group)? Organizational or institutional (large companies, villages or communities, for example)? Regional? National? Comparative across nations? Often, the unit of analysis will seem "natural" to the evaluator. Not much time, if any, will be spent debating the proper unit of analysis in most cases. However, sometimes it will pay to consider alternative units of analysis. A danger to beware, however, is the possibility of "ecological fallacy"; that is, gathering data, say, on the group level, then using that data to make statements about individuals within those groups. If data have been collected on the group level, the evaluator should confine analysis and conclusions to the groups and should not jump up or down the various units (Babbie, 1986).

Whatever approaches, units, etc. we use, the related concerns of measurement reliability and validity must be kept in the forefront of our consciousness. The universality of these concerns can be verified by picking up any textbook which discusses evaluation research and checking for a discussion of reliability and validity. For example, Smith and Glass (1987) repeatedly write about these measurement concerns, noting that if a measure is not reliable, it cannot be depended upon—it is inconsistent, imprecise, unstable. Reliability is necessary, but is not sufficient, for validity of measurement. Validity relates the meaning of a measure and the inference from that measure (usually a quantitative score) to the concept/construct which the evaluator intended to measure. Many ways have been designed to permit evaluators to check the reliabilities and validity of their measurements and probably are familiar to the reader. But let's not let familiarity breed contempt! If we do not have reliability and validity in our research evaluations, we have nothing. Every evaluation report should explicitly address these twin concerns without fail.

Bonus W: For Whom should Evaluation be Done?

An early author on evaluative research, Suchman (1967) insisted that evaluative research should be scientific (and largely quasi-experimental); however, as we shall see, this requirement is being challenged (Hornik, 1980). Patton (1980), a most influential writer in advocating that other social scientific methods ought to be considered and used in conducting research, said:

> ".... today's evaluator may be called on to use any and all social science research methods, including analyses of quantitative data, questionnaire results, secondary data analyses, cost benefit and cost effectiveness analyses, standardized tests, experimental designs, unobtrusive measures, participant observation, and indepth interviewing" (p. 18).

But even this array of social science methods has been criticized as being too restrictive. Alternative methods of evaluation have been sought in disciplines far from the social sciences. As metaphors, these methods may have usefulness in situations which are not amenable to the methods of Suchman and Patton. They include investigative reporting from journalism, law, architecture, geography, philosophy, literary and film criticism, and watercolour painting (Smith, 1981)! Also, possibilities may be found by examining the methods of folksongs, storytelling, still photography, research briefs, graphic display procedures, stem-and-leaf displays, operational network displays, geographic displays, oral policy briefings, briefing panel presentations, adversary hearings, committee hearings, and televisions presentations (Smith, 1982). The actual useful of many of these methods may be extremely limited; nonetheless, that they have been seriously proposed suggests that evaluators are in a mood to be adaptive and have been dissatisfied with traditional methods. There is no agreement on "the right method". Most evaluators who struggle with methods would find value in choosing several appropriate methods for a particular problem. An evaluator who uses the same method over and over again just isn't thinking very deeply about what he or she is doing. Also, there are obvious benefits to be drawn from using more than one method to study the same problem or phenomenon. Webb *et al.* (1966) and Denzin (1970) referred to this as "triangulation". Particular methods may have subtle biases inherent in their design and application; that is, they may tend to yield erroneous information which is dependent on the method itself and which does not achieve verisimilitude to the reality which the method/data are attempting to reconstruct or depict. Thus, triangulation. If different methods yield similar data pictures of the phenomenon of interest, the evaluator may have more confidence in the findings.

But these are niceties and subtleties which may concern only those with an academic and theoretical/methodological interest in evaluation. It is true that many evaluations seem to be conducted by academics who are primarily trained to communicate findings to other academics (Davis and Salasin, 1983; Weiss, 1972). At least some of this interest is due to the need to publish scholarly articles in order to achieve tenure and promotion. That's not all bad. The technical standard of research conducted for such purposes is liable to be high.

There are others for whom evaluation should be conducted and who should learn of evaluation results.

The most obvious group is made up of individuals who are responsible for planning and implementing the particular project or programme, as mentioned previously in the context of setting goals and objectives and in formative evaluation efforts.

Sponsors of the projects, programmes and evaluation — governments, organizations, groups, etc. — have an obvious right to data and results.

Those who conduct similar projects and programmes to the one being evaluated will naturally wish to be informed or to have access to information.

Other "stakeholders" include the general public, particularly taxpayers (if national money has been invested), students, the public service and business communities, etc. (Herman *et al.*, 1987).

Concern is sometimes expressed that no one cares about evaluation once it's conducted and reported. Reports gather dust on forgotten shelves. Recommendations lie fallow because they are not "on the agenda". It would be easy to become cynical about evaluation if one asked the question "What difference did the evaluation make"? Too often, evaluations make not the slightest whit of difference. Who wants to be part of such a morbid, funereal occupation? Not many evaluators, I would guess. But it is surprising, then, when we realize how passive most evaluators are and how unconcerned they are when it comes to the usefulness and impact of their work. Without being obnoxious or overly aggressive, I would suggest that evaluators ought to do whatever they can to ensure that their reports are read, their recommendations understood, and that action be taken on at least some of their suggestions. Otherwise, what good is it, other than providing financial support for the evaluator, and for filling up his or her hours, days, weeks, months, years with the same futile activities?

But, in keeping with the "participatory" research strategy mentioned earlier, please don't let us forget that the most important stakeholders are those ordinary and extraordinary persons in our midst who are most affected by attempts to plan social change. These people will include farmers in the fields, mothers with sick children, outcasts who suffer because of disease or bad luck of birth into a low caste, and all the others for whom many labour in order to try to improve their quality of life.

References

Agrawal, Binod C. (Ed.) *Anthropological Methods for Communication Research: Experiences and Encounters During SITE*. New Delhi: Concept Publishing Company, 1985.

Anderson, James A. *Communication Research: Issues and Methods*. New York: McGraw-Hill Book Company, 1987.

Babad, Elisha Y., Max Birnbaum, and Kenneth D. Benne. *The Social Self; Group Influences on Personal Identity*. Beverly Hills, CA.: Sage Publications, 1983.

Babbie, Earl. *The Practice of Social Research*. Belmont, CA.: Wadsworth Publishing Company, fourth edition, 1986.

Britt, Steuart Henderson. *Psychological Principles of Marketing and Consumer Behavior*. Lexington, MA: Lexington Books, 1978.

Campbell, Donald T. and Julian C. Stanley. *Experimental and Quasi-Experimental Designs for Research*. Chicago: Rand McNally College Publishing Company, 1963.

Campbell, J. Gabriel, Ramesh Shrestha, and Linda Stone. *The Use and Misuse of Social Science Research in Nepal*. Kathmandu: HMG Press, 1979.

Christians, Clifford G. and James W. Carey. "The Logic and Aims of Qualitative Research", pp. 354-374 in Guido H. Stempel III and Bruce H. Westley, (Eds.), *Research Methods in Mass Communication*. Englewood Cliffs, NJ.: Prentice-Hall, Inc., second edition, 1989.

Cronbach, Lee J., Sueann Robinson Ambron, Sanford M. Dornbusch, Robert D. Hess, Robert C. Hornik, D. C. Phillips, Decker F. Walker, and Stephen S. Weiner. *Toward Reform of Program Evaluation*. San Francisco: Jossey-Bass Publishers, 1980.

Cohen, Y. "Some Critical Reflections on Participatory Research". *Convergence* 15 (1982), pp. 77-80.

Davis, Howard R. and Susan E. Salasin. "The Utilization of Evaluation", pp. 381-427 in Elmer L. Struening and Marilynn B. Brewer, (Eds.), *Handbook of Evaluation Research*. Beverly Hills, CA.: Sage Publications, 1983.

Denzon, Norman K. *The Research Act: A Theoretical Introduction to Sociological Methods*. Chicago: Aldine Publishing Company, 1970.

Dissanayake, Wimal. "The Guiding Image in Indian Culture and Its Implications for Communication", pp. 151-160 in D. Lawrence Kincaid, (Ed.), *Communication Theory: Eastern and Western Perspectives*. San Diego, CA.: Academic Press, Inc. 1987.

Donaghy, William C. "Nonverbal Communication Measurement", pp. 296-332 in Philip Emmert and Larry L. Barker, (Eds), *Measurement of Communication Behavior*. New York: Longman, 1989.

Doshi, J.K. and Binod C. Agrawal, "Audience Profiles and Evaluation Research", pp. 217-224 in Ronald E. Ostman, (Ed.), *Communication and Indian Agriculture*. New Delhi: Sage Publications, 1989.

Edwards, Ward, Marcia Guttentag, and Kurt Snapper. "A Decision-Theoretic Approach to Evaluation Research", pp. 139-184 in Elmer L. Struening and Marilynn B. Brewer, (Eds.), *Handbook of Evaluation Research*. Beverly Hills, CA.: Sage Publications, 1983.

Emmert, Victoria J. Lukasko. "Interaction Analysis", pp. 218-248 in Phillip Emmert and Larry L. Barker, (Eds.), *Measurement of Communication Behavior*. New York: Longman, 1989.

Fortner, Robert S. and Clifford G. Christians. "Separating Wheat from Chaff in Qualitative Studies", pp. 375-387 in Guido H. Stempel III and Bruce H. Westley, (Eds.), *Research Methods in Mass Communication*. Englewood Cliffs, NJ.: Prentice-Hall, Inc., second edition, 1989.

Fitz-Gibbon, Carol Taylor and Lynn Lyons Morris. *How to Design a Program Evaluation*. Newbury Park, CA.: Sage Publications, 1987, 1978.

Fitz-Gibbon, Carol Taylor and Lynn Lyons Morris. *How to Analyze Data*. Newbury Park, CA.: Sage Publications, 1987.

Fitz-Gibbon, Carol Taylor and Lynn Lyons Morris. *How to Calculate Statistics*. Beverly Hills, CA.: Sage Publications, 1978.

Frerichs, Ralph R. and Khin Tar Tar. "Computer-assisted Rapid Surveys in Developing Countries". *Public Health Reports* 104:1 (Jan./Feb., 1989), pp. 14-23.

Haaland, Ane. *Protesting Communication Materials*. Rangoon: UNICEF, 1984.

Henerson, Marlene E., Lynn Lyons Morris, and Carol Taylor Fitz-Gibbon. *How to Measure Attitudes*. Newbury Park, CA.: Sage Publications, 1987, 1978.

Herman, Joan L., Lynn Lyons Morris, and Carol Taylor Fitz-Gibbon. *Evaluator's Handbook*. Newbury Park, CA.: Sage Publications, 1987.

Holsti, Ole R. *Content Analysis for the Social Sciences and Humanities*. Reading, MA.: Addison-Wesley Publishing Company, 1969.

Hoover, Kenneth R. *The Elements of Social Scientific Thinking*. New York: St. Martin's Press, fourth edition, 1988.

Hornik, Robert C. "Shedding Some Light on Evaluation's Myths", pp. 1-4, 15 in *Development Communication Report* 29. Washington, D.C.: Clearinghouse on Development Communication, U.S. Agency for International Development, January, 1980.

Houston, Tom R., Jr. "The Behavioral Sciences Impact-Effectiveness Model", pp. 51-65 in Peter H. Rossi and Walter Williams, (Eds.), *Evaluating Social Programs: Theory, Practice and Politics*. New York: Seminar Press, 1972.

Hyman, H. H. *Secondary Analysis of Sample Surveys*. New York: John Wiley, 1972.

Joshi, S. R. and Hansa Joshi, "Women and Television: ISRO Experiences", pp. 22-33 in Binod C. Agrawal, S. R. Joshi, and Arbind Sinha, (Eds.), *Communication Research for Development: The ISRO Experience*. New Delhi: Concept Publishing Company, 1986.

King, Jean A., Lynn Lyons Morris, and Carol Taylor Fitz-Gibbon. *How to Assess Program Implementation*. Newbury Park, CA.: Sage Publications, 1987.

Kish, Leslie. *Survey Sampling*. New York: John Wiley, 1965.

Markus, Gregory B. *Analyzing Panel Data.* Beverly Hills, CA.: Sage Publications, 1979.

McPhee, Robert D. and Donald P. Cushman. "Attitudes, Behaviors, and Messages: An Introductory Overview", pp. 1-41 in Donald P. Cushman and Robert D. McPhee, (Eds.), *Message-Attitude=Behavior Relationship: Theory. Methodology, and Application.* New York: Academic Press, 1980.

Morris, Lynn Lyons and Carol Taylor Fitz-Gibbon. *Evaluator's Handbook.* Beverly Hills, CA.: Sage Publications, 1978a.

Morris, Lynn Lyons and Carol Taylor Fitz-Gibbon. *How to Deal With Goals and Objectives.* Beverly Hills, CA.: Sage Publications, 1978b.

Morris, Lynn Lyons and Carol Taylor Fitz-Biggon. *How to Measure Program Implementation.* Beverly Hills, CA.: Sage Publications, 1978c.

Morris, Lynn Lyons and Carol Taylor Fitz-Gibbon. *How to Measure Achievement.* Beverly Hills, CA.: Sage Publications, 1978d.

Morris, Lynn Lyons and Carol Taylor Fitz-Gibbon. *How to Present an Evaluation Report.* Beverly Hills, CA.: Sage Publications, 1978e.

Morris, Lynn Lyons, Carol Taylor Fitz-Gibbon, and Marie E. Freeman. *How to Communicate Evaluation Findings.* Newbury Park, CA.: Sage Publications, 1987.

Morris, Lynn Lyons, Carol Taylor Fitz-Gibbon, and Elaine Lindheim. *How to Measure Performance and Use Tests.* Newbury Park, CA.: Sage Publications, 1987.

Hornik, Robert C. "Shedding Some Light on Evaluation's Myths", pp. 1-4, 15 in *Development Communication Report* 29. Washington, D.C.: Clearinghouse on Development Communication, U.S. Agency for International Development, January, 1980.

Houston, Tom R., Jr. "The Behavioral Sciences Impact-Effectiveness Model", pp. 51-65 in Peter H. Rossi and Walter Williams, (Eds.), *Evaluating Social Programs: Theory, Practice, and Politics.* New York: Seminar Press 1972.

Hyman, H. H. *Secondary Analysis of Sample Surveys.* New York: John Wiley, 1972.

Joshi, S. R. and Hansa Joshi, "Women and Television" ISRO Experiences", pp. 22-33 in Binod C. Agrawal, S. R. Joshi, and Arbind Sinha, (Eds.), *Communication Research for Development: The ISRO Experience.* New Delhi: Concept Publishing Company, 1986.

King, Jean A., Lynn Lyons Morris, and Carol Taylor Fitz-Gibbon. *How to Assess Program Implementation.* Newbury Park, CA.: Sage Publications, 1987.

Kish, Leslie. *Survey Sampling*. New York: John Wiley, 1965.

Markus, Gregory B. *Analyzing Panel Data.* Beverly Hills, CA.: Sage Publications, 1979.

McPhee, Robert D. and Donald P. Cushman. "Attitudes, Behaviors, and Messages: An Introductory Overview", pp. 1-41 in Donald P. Cushman and Robert D. McPhee, (Eds.), *Message-Attitude=Behavior Relationship: Theory, Methodology, and Application.* New York: Academic Press, 1980.

Morris, Lynn Lyons and Carol Taylor Fitz-Gibbon. *Evaluator's Handbook* Beverly Hills, CA.: Sage Publications, 1978a.

Morris, Lynn Lyons and Carol Taylor Fitz-Gibbon. *How to Deal With Goals and Objectives.* Beverly Hills, CA.: Sage Publications, 1978b.

Morris, Lynn Lyons, Carol Taylor Fitz-Gibbon. *How to Measure Program Implementation.* Beverly Hills, CA.: Sage Publications, 1978c.

Morris, Lynn Lyons, Carol Taylor Fitz-Gibbon. *How to Measure Achievement.* Beverly Hills, CA.: Sage Publications, 1978d.

Morris, Lynn Lyons, Carol Taylor Fitz-Gibbon. *How to Present an Evaluation Report.* Beverly Hills, CA.: Sage Publications, 1978e.

Morris, Lynn Lyons, Carol Taylor Fitz-Gibbon, and Marie E. Freeman. *How to Communicate Evaluation Findings.* Newbury Park, CA.: Sage Publications, 1987.

Morris, Lynn Lyons, Carol Taylor Fitz-Gibbon, and Elaine Lindheim. *How to Measure Performance and Use Tests.* Newbury Park, CA.: Sage Publications, 1987.

Neuber, K. A.. W. T. Atkins, J. A. Jacobson, and N. A. Reutermann. *Needs Assessment: A Model for Community Planning.* Newbury Park, CA.: Sage Publications, 1980.

Nunnally, Jum C. "The Study of Change in Evaluation Research: Principles Concerning Measurement, Experimental Design, and Analysis", pp. 231-272 in Elmer L. Struening and Marilynn B. Brewer, (Eds.), *Handbook of Evaluation Research.* Beverly Hills, CA.: Sage Publications, 1983.

Ostman, Ronald E. and J. K. Doshi. "Communication Research in Education and Training for Development", pp. 33-63 in Ronald E. Ostman, (Ed.), *Communication and Indian Agriculture.* New Delhi: Sage Publications, 1989.

Ostman, Ronald E., Sushma Iyengar, and Brijmohan Kothari. " A Focus Group: Key Informant Approach to Agricultural Communication Research", pp. 200-216 in Ronald E. Ostman, (Ed.), *Communication and Indian Agriculture.* New Delhi: Sage Publications, 1989.

Ostman, Ronald E. and M. C. Sekhar. "Models, Measurement and Multivariate Analysis: Qualitative and Quantitative Research", pp. 238-251 in Ronald E. Ostman, (Ed.), *Communication and Indian Agriculture.* New Delhi: Sage Publications, 1989.

Patton, Michael Quinn *How to Use Qualitative Methods in Evaluation.* Newbury Park, CA.: Sage Publications, 1987.

Patton, Michael Quinn, *Qualitative Evaluation Methods.* Beverly Hills, CA.: Sage Publications, 1980.

Reynolds, Paul Davidson. *A Primer in Theory Construction*. Indianapolis: The Bobbs-Merrill Company, Inc. 1971.

Riccillo, Samuel C. "Physiological Measurement", pp. 267-295 in Philip Emmert and Larry L. Barker (Eds.), *Measurement of Communication Behavior*. New York: Longman, 1989.

Scriven, Michael. "Prose and cons about goal-free evaluation". *Evaluation Comment* 3 (1972), pp. 1-7.

Smith, Mary John. *Contemporary Communication Research Methods*. Belmont, CA.: Wadsworth Publishing Company, 1988.

Smith, Mary Lee and Gene V. Glass. *Research and Evaluation in Education and the Social Sciences*. Englewood Cliffs, NJ: Prentice-Hall, Inc., 1987.

Smith, Maryann Yodelis. "The Method of History", pp. 316-330 in Guido H. Stempel III and Bruce H. Westley, (Eds.), *Research Methods in Mass Communication*. Englewood Cliffs, NJ: Prentice-Hall, Inc., second edition, 1989.

Smith, Nick L., (Ed.,), *Metaphors for Evaluation: Sources of New Methods*. Beverly Hills, CA.: Sage Publications, 1981.

Smith, Nick L., (Ed.), *Communication Strategies in Evaluation*. Beverly Hills, CA.: Sage Publications, 1982.

Solomon, Richard L. "An Extension of Control Group Design". *Psychological Bulletin* 46:2 (1949), pp. 137-150.

Stanley, Julian C. "Controlled Field Experiments as a Model for Evaluation", pp. 67-71 in Peter H. Rossi and Walter Williams, (Eds.), *Evaluating Social Programs: Theory, Practice, and Politics*. New York: Seminar Press, 1972.

Stecher, Brian M. and W. Alan Davis. *How to Focus an Evaluation*. Newbury Park, CA.: Sage Publications, 1987.

Suchman, Edward A. *Evaluative Research: Principles and Practice in Public Service & Social Action Programs*. New York: Russell Sage Foundation, 1967.

Tandon, Rajesh. "Social Transformation and Participatory Research". *Convergence* 21:2-3 (1988), pp. 5-18.

Triandis, Harry C. *Attitude and Attitude Change*. New York: John Wiley & Sons, 1971.

Yarmol-Franko, Karen. "Editorial Introduction: Focus on Participatory Research". *Convergence* 21:2-3 (1988), pp. 3-4.

Warwick, Donald P. and Charles A. Lininger. *The Sample Survey: Theory and Practice*. New York: McGraw-Hill Book Company, 1975.

Webb, Eugene J., Donald T. Campbell, Richard D. Schwartz, and Lee Sechrest, *Unobtrusive Measures: Nonreactive Research in the Social Sciences*. Chicago: Rand McNally College Publishing Company, 1966.

Weiss, Carol H. *Evaluation Research: Methods for Assessing Program Effectiveness*. Englewood Cliffs, NJ: Prentice-Hall, Inc., 1972.

8

Educating Farmers Through Better Communication

S.L. Ghosal

Within the past three decades or so there has been a burst of intellectual and practical activity in the area called, communication. Today the interested laymen and the sophisticated scholar are both concerned with communication; indeed, more and more of them from all fields are trying to integrate the contributions of communication theorists and practitioners with their specific disciplines.

Specialists and social scientists have defined communication in a number of ways but a significant point in all these definitions is that communication is a vital aspect of social change. The issue has assumed considerable importance under Indian situation because of heavy public investment in agriculture and rural development programmes since independence. Indeed there is need to evolve a dynamic and forward looking communication strategy to reinforce extension efforts of some one lakh extension workers working today in 17 major States under the Training and Visit extension programmes.

Recent Recommendations

Communication experts have in the recent past highlighted the need for identifying the critical areas that deserve concentrated efforts in terms of communication support; for example, how contact-farmers' efficiency could be increased. There have been recommendations for exercising utmost caution in ensuring consistency in message designs, whatever be the media used. Experts have also discouraged a flat approach to communication with a cosmopolitan target group; instead, specific target groups should be identified for specific message designs and types of media selected both according to the target groups and the message content, they have observed. And, above all, experts have been advocating a multi-media approach to help the farmers in the innovative decision-making process.

Communication support to extension, experts have given us to understand, is the coordinated use of different communication methods for the purpose of focussing attention on, and offering support to, specific areas of extension activities. The effort should lead the target group to retain the message and motivate action and to achieve this the communication strategy has to be carefully planned, communication specialists observe. The implications of all these recommendations in the communication process and their effects on other related factors perhaps deserve an analysis in detail.

Imperatives of Social Justice

The real test of success of communication process should however be measured by the speed with which the farmers are enabled by extension workers to adopt better farm technology. And herein lies the crux of the problem; for, farming cannot merely be dealt with only as technology or economics without consideration of its sociological and psychological implications. Through thousands of years the country has developed certain broad social, religious and philosophical customs and beliefs that gained universal acceptance and circumscribed the lives of untold millions of Indians—the rural folks in particular. In fact agriculture in rural India has for centuries been a highly integrated way of life involving family ties, religious sentiment and social intercourse. Any agricultural innovation that is recommended must be considered not only in terms of its expected productivity, but also in terms of what it would do to other parts of a farmer's life.

Treatment of Messages

Furthermore, recommended messages must however be presented verbally, visually and conceptually in such a way that the farmer perceives the potential advantages in much the same way as does the extension worker. And for this the contents of the message have to be decoded in an understandable language and treated in accordance with the level of comprehension and receptivity of the farmer. Proper treatment of the message also implies tailoring the message to make it commensurate with the resource capability of small and marginal farmers. In fact some of the major achievements of extension, under the training and visit system, may even result in, for example, a reduced application of fertilizer through recommending reduced, though more timely, application.

Of the four elements in a communicative act—source, message, channel and receiver—channel plays a central part. An important strategy for any communicator is to determine what channel to use in order to affect the receiver's knowledge, attitudes and behaviour. He cannot, for example, afford to use the powerful weapon of mass communication without an eye on his audience. In a country with a diversity of audience and conditions—and a scarcity of sufficient knowledge about them—if the communication media is to be useful, it must be based on facts rather than hunches and intuition, it must be adequate to allow for differences—cultural, social and ethical—among the various sections of the people and for changes with time. The problem is, however, that of getting a quick audience response to a particular media.

An Experiment on Audience Response

The Directorate of Extension of the Central Ministry of Agriculture tried to study the problem of feed back in terms of audience response in collaboration with University of Reading, U.K., in an area near Udaipur, Rajasthan a few years back. The idea was to check on an audience's response, under actual village conditions, to media and then feed the responses immediately back to the media producers.

In the three week old national workshop a group of social scientists and behavioural experts and another group of media men with a very high degree of expertise were assembled. It was decided to examine the response to one sample of each of the existing media type—poster, film, leaflet and radio programme—related to three matters of concern to farmers in the area: rat control, foliar spraying of urea solution on wheat crop, and poultry farming. The sociologists devised a questionnaire to reveal the degree of media comprehension. From the reaction samples drawn from some 80 farmers the first discovery was that majority of them did not receive the message at all. A drawing of a cyanide pump, for example, was taken to be a bicycle pump, and the rat was not understood to be dead. In a beautifully made film about foliar spray of fertilizer, the process was mostly mistaken for pest control. Meanwhile, in a KAP (Knowledge, Attitude and Practice) study with another group of farmers on poultry keeping a submerged layer of resistance was uncovered—chicken farming was considered not desirable because of the local religious prejudice against meat-eating. After it had become clearer where the weaknesses in communication lay fresh media materials were produced projecting the simplest messages as directly as possible. And an analysis of the newly produced material showed much greater comprehension and brought home the point that the more simple and direct the message, the more relevant the message with cultural norms, the greater was the degree of understanding and convictions that it carried. The study clearly reiterates the stand of eminent social scientists that we must exercise utmost caution while using the dangerous tools of mass communication.

Communication Techniques to Support Extension

Effective communication depends on both a suitable structure and appropriate techniques. The training and visit system of agricultural extension establishes a structure to facilitate communication in the linkage of extension, research, and farmers, created through planning and training sessions and the frequent, regular field contact between farmers and extension. To complement this structure, appropriate techniques for communication within the extension service, and between extension and farmers, must be identified and used. A number of communication techniques have proved effective in the training of extension staff, and also in convincing farmers to adopt production recommendations. To be effective, communication techniques used by extension must stimulate extension staff and farmers to become receptive to new ideas, and encourage feedback from farmers to extension staff and finally to research scientists. Five communication techniques have proved to be particularly effective in training extension staff: practical orientation, skill teaching, trainee involvement, samples and examples, and visual aids.

But techniques to enhance communication between extension workers and farmer must be practical and be accompanied by the teaching of skills and use of relevant visual aids, samples, and examples. There are number of communication techniques that an extension worker is expected to use. He is not advised to ask a farmer to adopt a new recommended practice on all his land or even on a major part of it; instead, he should aim at getting a large number of farmers—contact or others—to try the practice on a small area, for example, perhaps as few rows of maize planted at the recommended spacing, or planting at field corner with a new seed variety. This approach minimizes risk, establishes a series of scattered "demonstration" plots, and provides extension and research with numerous plots of field experiments. A variety of communication techniques may be used in training extension staff and in extension's contacts with farmers. But the ultimate test of the suitability of communication technique is its impact on audience.

Future Projection

The entire agricultural extension strategy of the country is being made consistent with the fast changing socio-economic conditions crossing the barriers of illiteracy and reducing the gap between the richer and smaller farmers. The question of systematic flow of farm information to the farmers of all categories has assumed great importance in that context. This should generate a new thinking in the methodology of farm information communication with a view of finding out the right communication tool that may help the Indian farmers—particularly the smaller ones—in the innovative decision process.

References

Foster, George M.—"Traditional Cultures and the Impact of Technological change" New York: Harper & Row, 1962.

Mathur, J.C. & Paul Newrath—"An Indian Experiment in Farm Radion Forum". uneases, 1959.

Rogers, Everett M. & F. Flyod Shoemaker—"Communication of Innovations," London: Collier-Macmillan, 1971.

Schramm, Wilber—"Mass Media and National Development," Paris: UNESCO, 1964.

Benor, Daniel & Michael Baxter—Training and Visit Extension, Washington, D.C., The World Bank, 1984.

Sreekantiah, K.N.—Communication Support to Field Extension Activity—a paper presented in the National Workshop on April Communication, at Madurai (July 17, 1986.).

9

Extension Education and Communication in Agricultural and Rural Development

Njoku E. Awa

Extension and Communication: An Overview

The agricultural extension personnel of low-technology nations bear a heavy responsibility in the attempts of their countries to increase food production. They are the linkage system between their governments' aims and their clients' needs; and in theory they bridge the "knowledge gap" in agricultural research and everyday practice in rural areas. Though crucial in introducing innovations to their clientele, however, they have been the object of surprisingly few studies—a mere 8 out of 708 summarised by Rogers and Stanfield (1966) in the UN First Development Decade of the 1960s. Attention has instead focused on the farmers and peasants whose behaviour they seek to change. Yet as James Green (1961:1) indicates in a study of Pakistani agricultural agents, most extension agents lack agricultural skills even in their own areas of specialisation. The agents studied by Green and his associates were proficient in writing lesson plans and describing skills for specific agricultural practices but deficient in performing the skills they had enunciated. The explanation, Green says, was simple:

> [They] themselves had been trained in a lecture system and had never performed the skill, or had done so only once or twice. There were veterinarians who had never castrated a bull calf; animal husbandry men who had never culled poultry;

horticulturists who had never pruned a tree; extension-methods educators who had never organized any kind of activity, club or council in the village, etc.

This is but one of the problems accounting for extension's fallibility in the development enterprise. Others include (1) status differences between extension agents and their clients, (2) agents' inadequate knowledge of the communication process; that is, "how communication works"; (3) lack of interagency cooperation both in programme planning and in implementation; and (4) extension's general lack of interest in traditional media.

A brief review of these limiting factors and apparent constraints sheds some light on extension in low-technology nations. Before we start that review, however, we need a few definitions to work with. As used here, ***extension*** refers to a mechanism for linking knowledge producers (research scientists) with users (clients) and implies a sharing of ideas, experience, expectations, and aspirations. In this sense, extension is an educational activity involving mutual learning and mutual influence between the change agent and his or her client. Similarly, *communication*, often perceived as a process merely of informing clients about new technical practices and exhorting them to adopt such practices, actually implies sharing. It is to the roles of these twin concepts in promoting or impeding agricultural and rural development, and their interaction with a host of situational and systemic variables (development incentives and disincentives), that this chapter is devoted. I also offer suggestions for strengthening the extension system and building communication infrastructures that are sensitive to indigenous knowledge and thus more attuned to the information needs and cultural assumptions of rural populations.

Status Differences

The very existence of an agricultural extension service demonstrates that changes in agricultural and other practices cannot be achieved merely by government fiat or by the simple dissemination of research findings. Extension agents form a conduit for introducing innovations at the local level. Their success thus depends largely on their ability as communicators and is affected by several variables, some of them sociological, others psychological. One sociological factor that contributes to productive and congenial interaction is status congruency. For reasons inherent in their position and training, extension agents generally perceive a status difference between themselves and their rural clients. Their socialisation experience, acquired mainly from secondary and post-secondary schools, differs markedly from that of their clients, especially small and subsistence farmers. Extension agents are, in a sense, part of their nation's elite through their position, education, and mobility. Yet while their formal education confers high status upon them, it ignores the fact that most of the agents "are handicapped in their efforts to communicate with lower status people because 'getting their hands dirty' is beneath their dignity" (Whyte 1975:9).

Complicating the problems of "status gap" are psychological variables that potentially can impair the agent's ability to initiate and maintain a meaningful relationship with rural clients. The central variable is homophily, "the degree to which pairs of individuals who interact are similar in certain attributes, such as beliefs, education, social status, and the like" (Rogers 1983:18). Homophily leads to an increased interaction between any A and any B who perceive similarity in their orientations towards "social reality." All communication may not be motivated by perceived similarity in beliefs and values; nonetheless people "prefer to associate with those they assume or perceive to be like themselves, apart from how similar or dissimilar they actually may be, and they tend to assume similarity and agreement with those they prefer" (Barnlund 1968:87). Extension agents differ from many of their clients in beliefs about technology and its value and in general orientation towards their environment. Since the majority of extension agents are college graduates and from urban backgrounds, their interaction with peasants is often impeded by a lack of opinion congruence and similarity in their world views; in short, by lack of homophilous relationships.

These various factors lend an "intercultural" quality to extension-client transactions: the agent is as much a stranger to the client as the latter is to the agent. This kind of relationship and its impact on knowledge and information utilisation is explained by Nathan Caplan (1979) in the "Two-Communities" Theory (the theory is similar in many respects to C.P. Snow's description of the gap between the humanities and the hard sciences in *The Two Cultures*). According to the two-communities perspective, change agents and their clients live in two separate worlds, each characterised by its own unique language and value system. If the two communities are left to their own devices, they are likely to communicate at cross purposes. Little empirical work has been done to test the validity of the two-communities argument, but there is some evidence that nonutilisation of information from one community by another can be attributed to the cultural gap between them (Rich 1979).

The situation is particularly critical in those instances (and there are many) where the agent is assigned to work outside the area in which he/ she was raised, but with no formal training in the dialect and customs of the community to which he /she is posted. A study of Eastern Nigerian extension agents posted outside their own villages (Rogers 1969: 182) demonstrates the point: because he used "a different Ibo dialect," Rogers writes, "an Ibo extension worker in Eastern Nigeria, employed only about 70 miles from his home village, was forced to communicate with his clients in pidgin English, which only a few of his clients fully understood." In this case the agent used substandard English, which he perceived to be mutually intelligible, to bridge a language barrier. But all communication is not verbal/vocal; some messages are transmitted nonverbally. Nonverbal or "second-order messages" are, in Gregory Bateson's (1955) terms, metacommunicational, that is, they designate for receivers the context in which specific verbal and vocal cues are interpreted. Nonverbal cues are messages about messages. Because some of these cues are culture-specific, they are hard to decode without knowledge of the cultural context in which they are used.

To overcome these linguistic and status barriers to intercultural communication, change agents must receive training in personal transformation—a process that requires the individual to develop greater sensitivity to his or her own cultural blinders and their limiting effects on interaction with strangers (Awa 1983). According to William Gudykunst and Young Kim (1984:229), the "process of personal transformation from being a stranger to becoming intercultural is potentially a process of growth—growth beyond our original cultural conditioning toward the ideal of the *intercultural person.*" It requires mental flexibility and two basic recognitions: that in a shrinking world no culture is an island by itself, and that relationship with others is impeded by rigid adherence to the psychological parameters and social rituals of one's own culture. The tendency, especially in Third World countries, for extension agents to demonstrate their cultural and intellectual superiority to farmers through artifactual and proxemic codes is often difficult to resist. Wearing a suit to a village meeting where everyone else is in blue jeans, wrapper, or *dhoti* is, for example, an attempt to build an interactional wall rather than bridge a cultural gap between change agent and client. The same is true of a change agent who demands a special seat at a meeting where everyone else is squatting. Such barriers to effective communication between change agents and their clients seem to lend support and credence to the two-communities perspective.

Agents' Inadequate Knowledge of the Communication Process

Most, if not all, extension agents are subject-matter specialists—agronomists, animal scientists, entomologists, horticulturists, plant breeders, food scientists. Their appointment is based on competence in a particular area of inquiry as demonstrated by the quality of their diplomas. Success in college courses in, say, plant breeding and animal physiology is generally mistaken as evidence of the agents' knowledge of the communication process, presumably because college education implies competence in oral and written communication. This mistaken assumption, is shared by agents and employers alike. The result is that agents unfamiliar with the nature of human communication, especially the psychology of interpersonal encounters, are given an enormously complex communication task to perform. (Agents trained in American landgrant colleges may not fit this characterisation, but they are often elevated to executive positions at the headquarters of their ministries shortly after they are appointed.)

A comprehensive overview of the communication process, or how communication works, would require systematic analysis of the various dimensions and levels of communication and their interrelationships. Such work is beyond the scope of this chapter. It will suffice to sketch the principal assumptions and postulates of interpersonal communication, with comments about their implication for change agent-client transactions. The reason for concentrating on interpersonal communication is that change agent-client encounters are mostly interpersonal in nature.

The first assumption is that *interpersonal communication is transactional.* Previous conceptualisations of interpersonal communication as primarily an action-reaction process painted a picture of a passive audience that had no ideas to share with a source and responded only when stimulated. In suggesting interpersonal communication is transactional, we mean that it is a process with interdependent components, involving participants who "act and react as whole beings" (DeVito 1989:21). Communication involves reciprocal effects between source and receiver. It is not something that one person does to another, nor is it one-way traffic.

The second assumption is that *communication is a process;* it is not static. The implication is that human communication is dynamic, continuous, and ever-changing. "As communicators," says Roger Wilcox (1977:17) "our mood, intent, point of view, and mental alertness are constantly changing from day-to-day, hour-to-hour, almost from minute to minute." Similarly, both the words we use and the meanings assigned to them change from time to time and, sometimes, from place to place.

Finally, *communication is symbolic.* In human transactions we use sounds—words in the language system we have acquired—such as the sound "dog," to symbolise a physical being that is different in kind from the sound by which we refer to it. Language is itself a symbol system. Communication is, therefore, effective only to the extent that source and receiver share meanings. The closer our words adhere to our listener's symbolic world, the more of our message the listener will comprehend and retain.

In extension communication, the choice of technically inappropriate symbols can retard adoption of an innovation. An example of this type of communication barrier is offered by D.L. Umali and G. O. Feliciano (1964), who studied the adoption behaviour of 45 farm household heads in the Philippines. Thirty-three of the 45 respondents said their extension agent used technical language that was incomprehensible to them.

Three Transactional Postulates of Interpersonal Communication

Joseph DeVito (1989) discusses six postulates of interpersonal communication derived from Watzlawick, Beavin, and Jackson's analysis of the psychology of face-to-face encounters in *Pragmatics of Human Communication* (1967). Three of these postulates which are pertinent to this analysis speak for themselves, but I add a word or two to put them in perspective:

(a) Communication is Inevitable

Communication is sometimes purposeful and intentional, but in many instances it is not. Often we think of communication basically in terms of verbal behaviour. But communication involves verbal and nonverbal behaviour, and in many cases the latter dominates interaction. And when there is conflict in meaning between the two channels, nonverbal codes assume dominance in the meanings that receivers assign to message cues. Except in Japan, most cultures socialise their members to believe that silence is not communication. But as Watziawick and his associates have found, all behaviour, including silence, is communicative. One cannot, therefore, not communicate. So all activity and inactivity, including words and silence, have a message content for the receiver; not, once enacted or said, can they be expunged from the head of the receiver.

This point is made most poignantly by John Gunther (1950:22) in a fascinating profile of the physical behaviour of former U. S. President, Franklin D. Roosevelt:

> In twenty minutes Mr. Roosevelt's features had expressed amazement, curiosity, mock alarm, genuine interest, worry, rhetorical playing for suspense, sympathy, decision, playfulness, dignity, and surpassing charm. Yet he *said* almost nothing.

Whether it be constant or erratic movement of our feet, a spasmodic contraction of our muscles, or unintended twinkles, our physical behaviours send numerous messages to people around us.

(b) Communication is Irreversible

Certain processes are reversible, others are not. We can, for example, turn water into ice and later reverse the process by allowing the ice to become warm. But this reversibility is not true of communication. Once we say something (e.g., laggards are lazy farmers who are unable to make rational decisions), it remains said. We cannot unsay it, no matter how much we try to qualify what we said. Understanding the significance of this postulate requires that we exercise caution and prudence in our choice of words when communicating with others, and that we guard our physical behaviour and the cues we send out through our facial and gestural behaviours to avoid sending wrong messages.

(c) Communication has Content and Relationship Dimension

Content refers to information conveyed to someone or in Bateson's (1962) term, the "report" aspect of a message. *Relationship,* on the other hand, refers to the way in which the source sees himself or herself in relation to the audience. Such relationship messages are expressed in one or several of the following assertions (Watzlawick *et al.*, 1967:52): "This is how I see myself......this is how I see you......this is how I see you seeing me," and so forth in hypothetically infinite regress. A relationship message is expressed in what Bateson describes as the "command" aspect of communication—"See me in my office, now." It is also expressed in tone (shouting, whispering). It is, in short, metacommunicational. To avoid confusion, it is important to realise that relationship messages are not always defined deliberately or sent with full awareness. In healthy relationships this aspect of communication tends to recede "into the background," whereas in sick relationships the content aspect of communication becomes progressively less important to interactors.

The primary implication of the content and relationship dimensions of interpersonal communication is in the management of conflict and its effective resolution. Often, failure to distinguish between the two levels of communication can create conflicts or worsen an existing conflict. It makes little difference who the interactors are—mother and daughter, husband and wife, extension agent and client—failure to recognise the distinction between content and the relationship dimensions of communication can cause conflicts.

Lack of Interagency Cooperation

Extension agents form but one link in what is best visualised as a chain of contacts and points at which information is disseminated in the process of technological change. They interact not only with their primary clientele—commercial farmers and peasants—but also with government agencies, which formulate national and regional policies in, say, agricultural development, and with the various research agencies responsible for developing cost-effective technical innovations. The extension agent's major function is to tailor these innovations into messages situationally relevant and appealing to his or her clientele, in order to ensure adoption.

When high-status development agencies, such as the agent's organisation, compete for recognition by higher-level decision makers and for scarce resources, the result is "noise" that can hinder or reduce the quality of communication between policy makers and extension agents. This noise can take the form of contradictory edicts and recommendations, high demand for paperwork to buttress cases in the competition for funds differing foci of interest, and so on. Such noise can hinder or destroy the efforts of the individual change agent even before he or she encounters problems in communicating with a rural clientele, causing frustration and misunderstanding about the agent's function.

Many agricultural programmes fail because they are not integrated with other development programmes to address the numerous related problems that face farmers. In a study of Nigerian extension agents, I found that organisations with goals similar to those of the Ministry of Agricultural Extension (namely, the Ministries of Adult Education, Community Development, Health, and Information) had no formalised, patterned interrelationships (Awa 1976). Some of the agents reported working with representatives from other Ministries, but these were temporary, working relationships established to concentrate on the solution of apparently transient, isolated problems and lasting only until the problems were resolved. Rather than pool resources in the formulation and execution of development projects, each agency functioned independently and reached the same audience with a slightly different message.

Lack of cooperation among organisations with similar or overlapping goals has resulted in the tardy recognition of the value of combining several factors to introduce *simultaneously* "a whole package of innovations, in order to maximize the effect of each" (Bunting 1971:442). Raising per-acre crop production while ignoring transportation can result in glutted local markets and lower prices for farmers who have extended themselves financially to purchase farm inputs; raising farmers' interests in fertilizer that is either unavailable or in short supply at the local level can result in frustration when the farmers are unable to obtain sufficient quantities to fill their needs. Until development agencies learn to recognise (1) the interdisciplinary nature of the problems facing their clients, (2) the need for a more flexible definition of their "areas of interest," and (3) the value of inter-agency cooperation, their contribution to the achievement of their country's overall development objectives will be minuscule.

Lack of Interest in Traditional Media

Extension agents concentrate on two forms of communication in the diffusion of innovations. The first is the medium of print, in the form of newsletters and leaflets, which are of little use in reaching the illiterate majority of their clientele. The second is oral or face-to-face contact. This latter form of communication seems to be ideally suited for the information needs of rural people. Unfortunately, oral channels of communication are susceptible to frequent misunderstandings through dialectal differences and variations in the use and interpretation of nonverbal codes. Furthermore, face-to-face contact with rural farmers is labour-intensive, and contact with clients with differential cultural norms is often stressful, in a psychological sense, for extension agents.

Because of their inability to fathom the local complexities of communication with their rural clients, extension agents and their ministries have tended to concentrate their efforts on the mass media. Except where a mass medium message has been adequately adapted to the needs and expectations of a rural audience, however, it has generally failed to address the sociocultural norms of such an audience; and proponents of change have often thrown up their hands when faced with an apparently incomprehensible melange of attitudes and beliefs, to concentrate their attention on those farmers most likely to adopt new ideas.

Why have extension agents refused to consult the potential strengths of traditional media in their communication with clients? First, the extension agent is in a position that is structurally cognate with that of traditional mediating communicators, such as the "gong-man" in Nigeria, the balladeer and mime in India, the folk singer in the Philippines, the puppeteer in Taiwan, or the towncrier in other societies; but he/she is separated by education and training from the culture of his/her clientele. The education of the agent concentrates on technical proficiency rather than rhetorical and persuasive skills, whereas the folk singer's or raconteur's role as an effective and credible communicator rests on a thorough grounding in the traditions of the village. Second, agents are generally taught to assume that media that have proven effective in reaching and convincing large numbers of people in the high-technology nations of the world are equally efficient in all societies. Experience suggests that this is not always the case.

One feature distinguishing traditional media is the opportunity they offer for close physical and psychological contact between message sources and receivers. Street theatre, for example, brings actors and spectators together to participate in a modified traditional play or indigenised Western drama. The value for development of traditional media lies in their spontaneous and improvisational character, which allows actors and raconteurs to alter message content to suit the needs and interests of a group. Whether it be storytelling, a puppet show, or a street play, the style is the same; that is, messages are composed extempore, the aim being to involve the audience as much as possible. In general, new twists are added to old plots, new characters are coopted as the need arises, and references are made to the current concerns of the community.

We must be aware, however, of cultural taboos that have rendered traditional media less democratic and thus less participative than they should be. For example, there are strong prejudices in India against women performing in street theatres. As research on persuasion and attitude change shows, beliefs deeply and fervently held by a culture are very hard to dislodge. But some women's reform movements in India have got around this taboo by organising street theatres and plays primarily for women. Traditional theatres and dramas should ideally involve men and women, especially when they are designed with social development in mind. Much research still needs to be done in this area, but the preliminary indications are that some folk media are better adapted to carry effectively the messages of change in which the future of low-technology nations is invested (Ugboajah 1979). To react sensitively to such opportunities, to tailor messages to suit the potential of those channels of communication which already exist, and to train extension agents to identify the opportunities for message dissemination—these eventually must be the aims of the modernising elites in low-technology nations.

References

Awa, Njoku E. 1976. "Communicating Agricultural Innovations in Nigeria: A Study of Government Change Agents". Paper presented at the 26th Annual Conference of the International Communication Association, Portland, Oregon, April 13-17, 1976.

——. 1983. "Culture and Credibility." *CERES*, FAO Review on Agriculture and Development, No. 95 (Vol 16, No 5), September - October.

Barnlund, Dean C. 1968. "Communicator Choice." In D.C. Barnlund, *Interpersonal Communication: Survey and Studies.* Boston: Houghton Mifflin.

Bateson, Gregory. 1955. "A Theory of Play and Fantasy." *Psychiatric Research* 2:39-51.

Bateson, Gregory. 1962. "Exchange of Information about Patterns of Human Behavior." Paper presented at a Symposium on Information Storage and Neural Control, Houston, Texas.

Bunting, A.H. 1971. "Agricultural Sciences." In J. Paul Leagans, and Charles P. Loomis, eds., *Behavioral Change in Agriculture.* Ithaca, New York: Cornell University Press.

Caplan, Nathan. 1979. "The Two-Communities Theory and Knowledge Utilization." *American Behavioral Scientist* 22, No. 3 (January/February): 459-470.

DeVito, Joseph A. 1989. *The Interpersonal Communication Book.* 5th ed. New York: Harper & Row.

Green, James. 1961. "Success and Failure in Technical Assistance." *Human Organization* 20.

Gudykunst, William B., and Young Yun Kim. 1984. *Communicating with Strangers: An Approach to Intercultural Communication.* Beverly Hills, California: Sage.

Gunther, John. 1950. *Roosevelt in Retrospect.* New York: Harper.

Rich, Robert F. 1979. "Systems of Analysis, Technology Assessment, and Bureaucratic Power." *American Behavioral Scientist* 22, No. 3. (January-February): 393-416.

Rogers, Everett M., and J. David Stanfield. 1966. "Adoption and Diffusion of New Products: Emerging Generalizations and Hypotheses." Paper presented at the Conference on the Application of Science to Marketing Management, Purdue University, Lafayette, Indiana, July.

Rogers, Everett M. with Lynn Svenning. 1969. *Modernization Among Peasants:* The Impact of Communication. New York: Holt, Rinehart and Winston.

Rogers, Everett M. 1983. *Diffusion of Innovations.* New York: The Free Press.

Ugboajah, Frank Okwu. 1979. "Developing Indigenous Communication in Nigeria." *Journal of Communication* 29, No. 4 (Autumn): 40-45.

Umali, D.L. and G.D. Feliciano. 1964. *A Socio-Economic Study of Selected Barrios in the Land Reform Area, Bulacao*. Philippines: University of the Philippines, College of Agriculture, Los Banos, Philippines.

Watzlawick, Paul, Janet Helmick Beavin, and Don D. Jackson. 1967. *Pragmatics of Human Communication: A Study of Interactional Patterns, Pathologies, and Paradoxes*. New York: Norton.

Whyte, William Foote. 1975. *Organizing for Agricultural Development*. New Brunswick, New Jersey: Transaction Books.

Wilcox, Roger P. 1977. *Communication at Work: Writing and Speaking*. Boston: Houghton Mifflin.

Part 3

Research-Extension-Farmers Linkages—A New Thrust for Development

"Extension must help bridge the social, cognitive and geographic distance between a small group of scientists and a large farming community by selecting or designing methods most appropriate for particular message content and the farmer's state of readiness...."

J. Lin Compton, 1984.

10

"We did it Ourselves" Farmers' Participation in Research and Extension

Laverne B. Forest

Introduction

"We did it ourselves" was the answer a southern India farmer gave as the central reason for success of an extension agricultural development project. "We have to rid ourselves of this socialist illness if we are to achieve agricultural development", was a criticism of central government planning, and the reason for lack of success given by a high placed government agricultural administrator in a communist ruled country.

These two contrasting quotes from two different countries both support the same premise of this chapter—Farmers and local rural people must actively and of their own choice, participate in the agricultural research, extension, and development program for their rural communities if development is to be achieved. This premise and its underlying philosophy are true in America where involvement of farmers and other clientele in agricultural extension programs have a long standing tradition, but also in Southeast Asian developing countries.

This chapter describes: (1) the concepts of farmer participation in research and extension, (2) the purposes and potential benefits of such practice, (3) developing countries study which supports further such participation, and (4) basic recommendations for involving farmers in rural communities in research and extension.

The Involvement Concept

Two terms, participation and involvement, are oftentimes used interchangeably, and this chapter will be no different. The term involvement perhaps might be used more frequently than the term participation. Both essentially mean the same thing. To begin the discussion then, the following definition is provided:

> Farmer involvement in extension and research is including them in the research and extension process by creating experiences in which they become, socially and psychologically, a part of that process, not just physically present (Forest, 1988).

We are discussing involvement here as Paul (1987) says, "an active process whereby beneficiaries influence the direction and execution of development projects rather than merely receive a share of project benefits" (p. 5). This article does not deal with wider community participation or the on-going spontaneous, indigenous, and local participation found in all communities.

Several terms need further definition. Involving leaders, farm youth, farmers, and other agricultural representatives, in the research and extension process includes a broad spectrum of degrees, attitudes, techniques. Gathering a small group of farmers in a village meeting place, is physical involvement. When they discuss their frustrations, concerns, needs, dreams, and future plans for their farm and for their farming village, that is social involvement. However, they can be involved even more. If that same hypothetical group of struggling, poor farmers, through active discussion and analysis led by extension agents and educators, internalizes certain research and extension alternatives, they could, in turn, act together to promote adoption of particular program plans.

When farmers identify their own problems, problems' causes, and needed research, and then relate the research to their own needs, the farmers are not only involved physically and socially, but also psychologically. This deeper psychological involvement and commitment can be a key goal in itself for any research and extension process. The deeper commitment has been shown to lead farmers to publicly and politically support extension. The commitment itself becomes a resource for conduct of the program.

Involving farmers assumes several key principles:

1. All farmers are always involved, if only indirectly in every research and extension situation. At the very least, their lives are a part of what extension agents do to them. The challenge is therefore not merely getting farmers involved, but having them involved for purposes of advancing agricultural development, extension, and research.
2. Involving farmers in research and extension demands special leadership and communication skills on the part of extensionists and researchers.
3. All efforts at publicly involving farmers must be open, honest and non-manipulative. Continued effective involvement leading to true expressions of need, trust, and support cannot be obtained through dishonest and manipulative approaches.
4. Neither ethics nor democracy require that all farmers be directly involved all of the time in every aspect of the extension and research process. Rather the aim is to involve them for specific purposes so that farmers will have influence on decisions affecting them.
5. Having the farmers make all the program decisions does not necessarily guarantee democracy. Total farmer control could possibly result in involving the "wrong people".
6. Involving farmers in the extension and research process helps identify the research needs and resources, but also generates support, learning, and knowledge of constraints.
7. Full psychological involvement and commitment of farmers allows various values and criteria for data interpretation to be discussed openly and in non-threatening ways. They all support use of varying personality approaches.

More will be said later in this chapter, but it is hypothetically possible for farmers to be involved in research, in extension, and a combination of the two. It is also hypothetically possible for farmers to be involved in the endeavors from the beginning to the end—that is, in the planning, implementation, and follow-up stages.

Finally it is hypothetically possible that farmers can be not only physically involved, socially involved, and (most importantly) psychologically involved in many ways. Involvement is therefore a broad concept and this chapter is in the context of this broad concept.

The Purposes and Potential Benefits of Farmer Involvement

Forest (1981) stated that goal setting should be done locally. The logic of this statement is 1) decisions on research and extension activities should be good decisions. 2) Good decisions are those which lead towards positive valuable programs for farmers. 3) Valuable programs are ones which are effective, relevant and efficient. 4) Effectiveness means these extension and research programs achieve positive results whether initially intended or not. These results are new knowledge, attitudes, skills, actual practice changes, and benefits from applying the new farming practices. 5) And finally these result more likely happen with involvement.

Bamburger (1986) raises several excellent considerations. He suggests that the medium or the method for involving farmers is important. He also suggests that attention be given to the stages of the project in which farmers are involved, the program level or scope, and the intensity of participation (information sharing, consultation, decision-making, and initiative taking). Professionals should consider the number of farmers involved, the duration or regularity of farmer involvement, and whether or not women who are playing large roles in agricultural work are involved in the process.

Adequate consideration to these questions will mean that Paul's (1987) proposed objectives to participation might be achieved: (1) sharing project costs, (2) increasing project efficiency, (3) increasing project effectiveness, (4) building beneficiary capacity, and (5) increasing empowerment.

The benefits and valuable programs have been shown to be more probable when people are locally involved. Thus purposes and benefits for involving farmers in the research and extension endeavors are crucial. They fall into five categories (Forest, 1981):*

1. Democratic ideals
2. More inclusive identification of needs and decision alternatives
3. Program acceptance and support
4. Learning benefits
5. Nature of reality

Democratic Ideals: People can and should decide their own destiny. Abolition of slavery, resistance to totalitarian governments, and demands for and enactment of civil rights legislations in all countries of

*See also Crouch, 1983, Cellman, 1987.

the world are ample proof that people believe in democracy and their rights and privilege to make decisions affecting their future lives. Recent trends in countries with totalitarian governmental systems add further proof, this is the era of more participation by people in decisions. Recent local, popular and open elections in such countries as Russia, Poland and Yugoslavia along with the pro-democracy demonstrations in China mean that the spirit lives, as past leaders such as Churchill, Jefferson, Lincoln, and Ghandi meant it to be.

More Adequate Needs Identification: Whale (1966) found that needs identification in decision making process was enhanced by local involvement of clients. Local clientele identified 41% unique needs not identified by professional educators. He replicated this experiment four times. Extension clientele do add to the quality of the needs identification and the goal setting process. They also validate the needs identified by the professionals.

Especially critical is the relationship, however, between needs identification and individual motivation. Any person regardless of level of education or culture knows their specific wants and orientations better than external professionals. Second, the voluntary nature of taking part in extension and research necessitate that farmers perceive the program as meeting their own needs. Under those conditions they will be more motivated to participate and learn in eventual programs. The closer the needs identification and goal setting process is to the farmers' farms and villages, the more likely will goals be specific and unique to their values and needs.

Gaining Program Acceptance and Support: Experiential philosophy (Dewey 1939), suggests that the value of a goal depends on having it evolve from that person's own experiences.

Second, psychological congruence theories suggests that if individuals believe in the need and goals they will also believe in the means toward resolving the need, i.e., participation in eventual programs, as well as the opposite, i.e., if individuals do not believe in the needs and the goals, they will also not believe in their means and methods to alleviating those needs.

Third, research on resistance to change suggests that resistance will be reduced if the farmers feel the project is their own and not one devised and operated by outsiders; if the project is in accord with their values and ideals; if farmers feel their security is not threatened; and if farmers have joined in diagnostic efforts leading them to agree on what the basic concern and problems are.

Fourth, adoption and diffusion research suggests that educational change strategies must consider the values, knowledge levels, power basis, resource and past experiences of the social systems within which potential participants perceive their needs.

Fifth, participation effects acceptance of goals (Forest 1973, 1978, Sail and Forest, 1985). They found that leaders and farmers overt observable commitments and involvements directly related to their acceptance of intended goals. Forest (1977) determined that perceptions of extension's effectiveness in worth in tax dollars related directly to rural leaders' amount of contact with extension agents through local committees set up by extension for planning (see Figure 1 illustrating the relationship between amount of committee involvement and the degree to which leaders perceive extension positively).

Fig. 1: Relationship of Leaders' Judgments of Extension's Worth in Tax Dollars and their Committee Contact with Local Agents

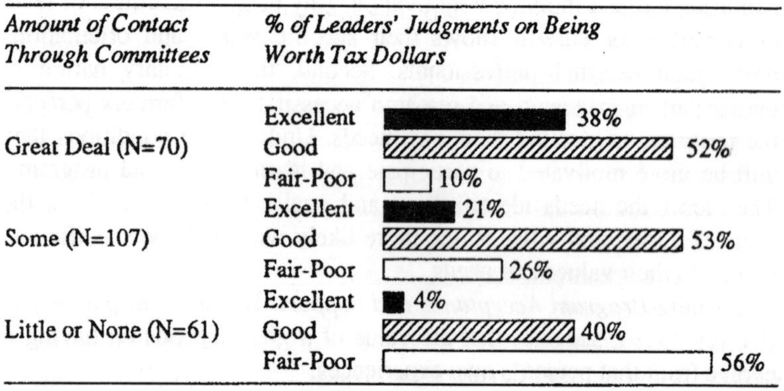

Reproduced from Forest, L.B., *Shawano County Leaders: Judgments About Extension, No. 9, 1977.* See also Forest, L.B., *Shawano County Leaders: Involvement in Extension Program Development, No. 6.*

Learning and Other Benefits: Farmers do not learn new technologies only from meetings planned for them by other people. They also learn new farming practices while planning future extension and research activities. By being part of planning and implementation meetings they have a chance to talk with other farmers, researchers, and extensionists. All of this interaction increases the chances for learning new ideas from neighbors, extensionists, and researchers. Learning is a continuous process. They learn more about their own situation, and its inadequacies, learn of others' experiences, identify realistic needs, and identify possible solutions.

Goal Setting Reality: According to Dewey (1939), any goal, predetermined by nonfarmers, is not really a set goal but is vulnerable to being made more specific and meaningful by the farmers who eventually participate in a related program. Individuals, determine the practical reality in the actual situation. Farmers determine for themselves, the importance of an extension or research program goal to their lives. They must understand the underlying reasons for the goal to be their reality.

To summarize two points are important. First, learners must be involved with extension and research. Cynics might easily argue that these are all very fine reasons for increasing the participation of farmers in research and extension but only in the United States or other developed countries. These cynics would suggest these are not reasons for applying participation to Indian culture or farmers or other southeast Asian developing countries. They could say farmers in developing countries are not prepared to participate, are not wanting to participate, and do not participate as adequately as the professional researchers and extension people. Yes, apathy might exist on the part of some. However, most of this apathy is due to tradition created by extension and research professionals, and other figures of authority (Ahmad, 1987). Farmers and rural leaders do learn to participate when given the opportunity. They must be given the opportunity by professionals who see the above potential benefits of active farmer participation on research and extension.

Second, as Compton (1989) points out, "It is fundamentally important to agricultural development to achieve effective interaction between research and extension." Involvement must reach all parts and all persons affected by agricultural development.

The Study of Good Agricultural Extension Programs

Forest (1989) focused on one question to selected persons in different countries: "What are the characteristics or ideals of a good agricultural extension program?" Four categories of people were interviewed: (1) Farmers who were extension clientele; (2) Local extension officers; (3) Extension administrators; (4) Community leaders. The countries for this study were Finland, United States, Yugoslavia, India, Nepal, Malaysia, and the Philippines. The interviews were about one hour long. Interviews were both individual, as well as small group interviews. The interviews were informal and non-leading. Follow-up or probe questions were asked only if an idea had already been mentioned. "Good" was not defined until interviewees offered their responses. They had to define success, achievement, change, development, reconstruction, improvement, etc. if they chose. In other words, all interviewees were left to define in their own ways what they valued, considered important, believed to be good about good programs, without intentional biases and definitions introduced by the author. Most of the interviews were in English with translation and interpretation provided when necessary.

The study's null hypothesis was—respondents will vary significantly in their response to the question, depending on their country, cultural system, extension system, and past experiences. It became obvious relatively soon in this study that the null hypothesis would be rejected. The data accumulation began to illustrate trends and themes, and these trends began to surface in every country visited. The following general themes were found to be inherently desired characteristics in good agricultural extension programs:

1. *Situational Relevance:* Client needs and situations serve as the basis for programs. Clients must have a readiness for the extension program. They must see the relevance of the program ideas to their own situation and needs. The program goals and the priorities must be important to them. Examples of quotes supporting *situational relevance as* a theme of good programs were:

— "Technology fits the farmers."
— "Match the Farmers wants and abilities."
— "Focus on real problems of local farmers."
— "If doesn't fit needs, not used."
— "Be aware and adapt to different needs as timing, season or time goes on, be sensitive to changing situation."
— "Communication with farmers—language must be appropriate."
— "Identify felt needs of people."
— "Actual need or interests of farmers—need based."
— "Involve and relate to local leaders, politicians, and traditional leaders."
— "Participation of people important."

2. *Research*: Unless Extension has knowledge and research to extend and give to farmers, the extension program cannot be a

good program. A research base (basic and applied) is essential regardless of existing farmer and village knowledge.

Statements supporting *research* as a key theme to good programs were:

— "Agent introduces new improved varieties with high yielding capacity."
— "Technology is more superior to local."
— "Provide technology."
— "Need variety tested in farmer's field."
— "Program is successful when it is technically sound,"
— "New technology."
— "Must be new material each time to increase participation."
— "Good research and knowledge needed."

3. *Resources and Inputs*: Program inputs and resources are budget, facilities, roads, access, media and other such items. However, in the case of many developing countries including India, it was found that resources and input include some provision for assistance on farmer input such as seeds, fertilizers, pest control measures, and marketing facilities.

People said *resources and inputs* were important by saying:

— "To arrange input management."
— "Allocation of budget must be top priority."
— "Gets inputs in time, timeliness."
— "Materials support."
— "Irrigation equipment."
— "Should be logistical support such as tools, resources, materials, to use right on the spot at the right time."

4. *Qualified Staff*: Extension staff must be qualified on the subject matters they are teaching, but also on their methods of teaching and relating to people. Staff need to be up-to-date on research, but also knowledgeable on farmer situations and needs if these extension people are to relate the new research to the farmer's situation.

Regarding *staff qualifications*, comments made included:

— "Extension agent should have enough knowledge on extension."
— "Need a teacher, a tailor master who can teach."
— "Both technology and extension methodology needed to run a good program."
— "Optimal level of training."
— "Be able to express ideas to communicate, to adapt."

5. *Good relationships* and cooperation must be observed between all agencies, leaders, communities, farmers, extensionists and researchers.

   Duplication, conflicts, and non-involvement of certain key people who can add resources, only leads to less effective and less valuable programs.

   Good *relationships* as a critical factor to good programs were illustrated by:

— "Confidence of farmers, good rapport needed, they need to become believers in program."
— "Be sensitive and aware."
— "Involves local officials because 60-70% know him."
— "Failures due to no sincerity."
— "Two-way communications needed."
— "Cooperation from all segments."
— "Programs need political support, relationships important."
— "We need trust in programs, relationships and support important."

6. *Good Methods and Media*: Extension staff must use demonstrations and other means which can be seen and participated in by the farmers. Verbal messages are not sufficient. Actual experiential teaching and research methods must be used if the programs are to be good.

   Illustrating *good methods and media* as key themes to good programs were these comments:

— "Share - tell - explain what farmers should do."
— "Demonstration sites on road sides."
— "Agents need to teach practically, so farmers can learn technique."
— "Make demonstration plot, take farmer tours to area."
— "Select farmers along well traveled road where other farmers will also see technology."
— "Field visits and field days must be regular."

7. *Program Results*: Results were almost assumed by the people as defining good. The program must lead to farmers knowing more, changing, and achieving ultimate benefits in their communities and on their farms. Learning and practice changes, and increased production should occur. In support of program results as a key ingredient to good programs were:

— "We need to develop this feeling of accountability among staff to farmers."

— "Impact is positive on farmers."
— "Program is adopted well."
— "Productivity improved."
— "If output of program more than local methods bring about change."
— "Program is successful when economically beneficial."
— "Solved problem."
— "Increased yield and benefit."
— "If no benefit, not a good program."
— "Need adoption and transfer of technology."

The above seven themes of good extension programs identified by interviewees are extremely important in several ways. First, each theme provides a reason for actively involving farmers in research and extension. Second, each provides suggestions on what professional extensionists and researchers need to do in order to effectively involve farmers. Specifics for each of the themes will help illustrate these points.

1. *Situational Relevance*: Each village and farm is different regardless of similarities. A researcher or extensionist cannot possibly know a farmer or village situation and their specific needs without interacting with them. Involvement of farmers must occur, and with a representative group of the various types and characteristics (farm size, location, gender, crop system).
2. *Research:* Basic research can occur in the laboratory, but applied research needs to fit the farmer's situation. Thus, various representative farmers must be involved to share (show and tell) what they believe would be useful and practical research.
3. *Resources and Inputs:* Involvement has been empirically shown to build and increase program resources. Village facilities, good will and trust, and informal communication are examples of increased resources because of effective involvement. Thus, open, honest observing, listening and interacting must occur.
4. *Qualified Staff:* Respondents stated extension staff needed to know processes of teaching, listening, sharing, relating and communicating which are all involvement oriented skills and abilities. Thus respondents were implying involvement is essential, and professionals have to be good at the skills.
5. *Good Relationships:* An extensionist or researcher cannot have good relationships unless they themselves interact and participate with leaders, farmers and all agencies. Sincere involvement develops good relationships with farmers.

6. *Methods:* Demonstrations and other effective teaching methods require involvement of the learner. Farmer learners do not learn just by listening passively to an instructor. Active 2-way interaction in meetings, workshops and classes is involvement and it must occur.
7. *Results:* As has already been stated, results and effects are important to good extension programs, and active involvement increases effects (i.e., learning, practice adoption, increased production, and rural changes). The logic is very clear—psychologically involve farmers in research and extension.

Perhaps the reader can detect that this study did not necessarily discover new ideas. Most of the ideas about farmer participation in extension and research have been considered and practiced in many countries of the world. However, this study does validate the process as feasible, useful, and practical for India and other Southeast Asian countries. The thesis of this chapter—Farmer Participation in Extension and Research as a viable practice—was supported by this study. The study did not refute that farmers can be or should be involved in the extension- research process. It did suggest ways or methods by which farmers could be involved, but further research is needed on that question also.

Based on the literature reviewed so far, and this study discussed, Table 1 summarizes the types of involvement in extension and research for (1) farmers and leaders, (2) extensionists, and (3) research scientists. In Figure 1, one dimension shows phases or stages of the agricultural extension program development process: (1) Planning, (2) Implementation, (3) Evaluation.

Table 1, also shows three functions in which the three categories of persons can potentially be involved: (1) Basic Research, (2) Applied Research, and (3) Extension.

None of the phases are distinctly separate from the others. In fact, the phases generally overlap one another with a good deal of planning occurring in implementation, evaluation, etc. Furthermore, the figure is not meant to suggest basic research, applied research and extension are separate from each other. These also go on simultaneously and in many cases one experience (e.g., field demonstrations can both create new applied research as well as extend ideas to farmers). Nevertheless, the presentation of these possibilities is useful here to show how different groups of the people can become involved in the practices of research and extension.

Table 1 shows that researchers, extensionists and farmers/leaders can and should be involved in all of the resulting nine cells of the matrix with only two exceptions. The two exceptions are in Cell 2 where basic research and implementation intercept. This is to suggest that both extensionists, and farmers, do not have the experiences, knowledge, skills and very likely the attitudes to implement basic research. Otherwise the table says these three categories of persons, regardless of agency of institution, can and should potentially be involved in the entire process. Involvement of the three can occur in almost all of the extension/research endeavor.

Table 1: Participation in Research and Extension

Program Function	*Program Phase* — *Planning*	*Implementation*	*Evaluation*
Basic Research	1 Researchers Extensionists Farmers/Leaders	2 Researchers	3 Researchers Extensionists Farmers/Leaders
Applied Research	4 Researchers Extensionists Farmers/Leaders	5 Researchers Extensionists Farmers/Leaders	6 Researchers Extensionists Farmers/Leaders
Extension Teaching	7 Researchers Extensionists Farmers/Leaders	8 Researchers Extensionists Farmers/Leaders	9 Researchers Extensionists Farmers/Leaders

To summarize to this point—to involve or not to involve farmers is not the question. Research, literature, and the 1988 study all unequivocally say they should be involved. The hypothesis is that researchers, extensionists, and farmers can and should be involved. The more crucial questions then are about the nature of the involvement, the roles performed by each person, the potential contributions of each group on the timing, and amount of involvement. The chapter now focuses on these more specific crucial questions of involving farmers, extensionists, and researchers in each of the nine cells in Table 1.

Cell 1—Planning Basic Research: Researchers obviously rely on their knowledge of the literature and past basic research, and state what research is feasible and should be done given the knowledge of research, techniques, data gathering, and data analysis. But extensionists, who are in the field with the farmers, and farmers themselves, contribute to planning by stating questions in need of answers. They can suggest concerns and express their beliefs about what would be helpful on their farm from their point of view. In suggesting needed basic research, farmers feel they belong to the program. Farmers also share their own personal knowledge of how they farm. This local knowledge is useful to the researcher as he continues to plan his specific basic research at the station or laboratory. A critical point here is that researchers decide what specific basic research will be done. They decide the goals, the priorities, the methods and the needed resources to conduct the basic research. Farmers and extensionists advise the researcher on goals and needs but do not necessarily decide.

Cell 2—Implementing Basic Research: This cell more than any other presents difficulties for collaboration, between the three categories. The researcher is trained in the methods, resource management, techniques, etc., any of necessity implements basic research. The farmers and extensionists might be interested bystanders. It is much more difficult for them to be meaningfully involved in this cell.

Cell 3—Evaluating Basic Research: Researchers compare validity and reliability of results with other basic research, and research peers judge the value by accepting or rejecting the new discoveries. Certainly researchers are involved in this activity. However, the ultimate value of new basic research findings is whether or not they lead to applied research or lead to actual application of new practices by farmers. Thus, extensionists and farmers inevitably are involved in judging or placing value on the basic research. Researchers must recognize this ultimate, unavoidable, practical judgment placed on all research by the farmers and village extension workers.

Cell 4—Planning Applied Research: All three categories of persons can be involved in planning results demonstrations, and techniques demonstrations. If the researcher have found value in listening to farmers on basic research, they'll definitely find value in listening to farmers and extensionists on needed applied research and how the research should fit in with local farmer knowledge and needs. The applied research should be done in the villages and on farm fields. Thus farmers and extensionists serving as the bridge between research and farmers, will of necessity be involved.

Cell 5—Implementing Applied Research: Researchers are involved to provide continuous guidance and rigor to how the research is done in the field. The extensionists are involved in conducting the applied research and the farmers are involved as observers, as workers, and as conductors of applied research in their local villages. They will feel involved if they in fact have had a real role in planning.

Cell 6—Evaluating Applied Research: Researchers are involved as they compare the applied research results to the basic research results as well as to applied research in other settings and in other villages. The researchers thus place value on the validity and reliability of the applied research. However, the extensionists and farmers provide the ultimate judgment about validity and reliability. They are the ones who will judge whether the applied research results and techniques are generalizable to fit local villages and are appropriate given other local knowledge existing in the farmers and villages. Extensionists and farmers are involved in this cell whether researchers and others plan for it or not.

Cell 7—Planning Extension: Typically in the past, extensionists have planned extension programs. However, the researchers have a great deal of knowledge to offer from their basic research and applied research in many villages. They should contribute as they are now doing to what are feasible and reasonable extension programs. Also, farmers have much to contribute; they have the problems. Farmers have the needs and the existing situational knowledge into which the new research and knowledge has to fit. Thus farmers have to be involved in providing the place and in this case deciding priorities and goals of future extension programs.

Cell 8—Implementing Extension: Researchers will not necessarily be as involved as extensionists or farmers in this cell. However, researchers are specialists and can be guests or can be speakers at local village meetings. Researchers can write materials that farmers can understand, use and refer to. The same materials will be interpreted by extensionists for further follow-up work with farmers. Extensionists are obviously involved in implementing the extension program but have to ensure materials are practical. Farmers are obviously involved as clientele or recipients of the new knowledge being advocated. However, farmers can be more than just recipients. Farmers can also be teachers of others. Farmers can implement results from their own farms through interaction and sharing with other farmers in local villages and on-site, results demonstrations. Farmers can be much more involved than just passive receivers of new ideas.

Cell 9—Evaluating Extension Programs: Researchers are involved because they are interested in having their basic research and applied research results from other villages adopted by farmers, the ultimate pragmatic value of their research. They collect and write about data on practice and adoptions. Extensionists gather data, monitor, and reflect on the amount of learning, practice adoption, and benefits due to extension programs. Farmers obviously are the key judge or evaluator. Farmers decide whether the extension programs have brought appropriate, feasible new knowledge to the villages and communities based on basic and applied research. They also compare and fit the new knowledge to the existing knowledges of the villages and farmers.

Techniques for Involving Farmers

In each of the nine cells in Table 1, farmers can be involved in research and extension in various ways: (1) Individually, (2) Groups, and (3) Existing data.

Involve Individual Farmers: One can conduct surveys by visiting each farmer (or a sample), and interviewing them. Surveys can collect data on their needs, their interests, current knowledge, actions, skills, changes and benefits due to Extension. Farmers can also be asked about their future goals, existing resources, needed resources, and barriers they believe are keeping them from improving their farm and their rural community. Keep the questions and surveys simple and short. Explain the purpose of the questions and the intended use of the survey. Listen to what they say, not to what one thinks they mean.

Involve Groups of Farmers and Villagers (and extensionists and researchers also): Group meetings and discussions can solicit clientele and farmer needs, knowledge levels, attitudes, skill levels, programs, and research preferences. Within group meetings, various techniques can be used to obtain farmer opinions. One technique is to brainstorm (Osborne, 1979). This process encourages open sharing of ideas with anyone speaking at any time.

A second group approach is the nominal groups techniques (Delbecq, 1975), modified to fit developing rural communities. The essential idea with nominal groups is to allow each individual time to think about their needs in planning, in implementing, and in evaluating programs. Once farmers have thought about what they want to say, the group leader goes around the group and asks each individual farmer for one idea until each has given each of their ideas. As these ideas are given, the leader records them on a sheet of paper or writing surface for the group to vote on later.

The group can also be interviewed as a group. The group can be asked a series of questions, giving each individual a chance to answer. This technique has been used by the author, and responses imply that if one or two people speak, others are supporting it unless they speak up and express a different point of view.

Formal hearings are another group method. Farmers and rural clientele can be involved by listening to a proposed program and asking them for their ideas on alternative needs, alternative goals, future priorities, and interpretations of what is needed in the rural village.

Leaders of extension programs and research can also help community leaders organize advisory councils, task forces, and committees. Developed countries have many years of successful experience in organizing these types of formal groups. Some of them can be sustained for a long time (advisory councils) and some might be for only the immediate short term task (ad hoc task forces).

Farmers can also be organized into small groups, with each small group looking at their needs and planning appropriate strategies according to their beliefs and feelings. Each group would then present their alternatives to another group. The presentations to each other and resulting interactions cause new ideas; the larger group can then agree on the priorities and needs. The smaller subgroups can be charged with advocating their plan and defending it. This act of defending will generate further commitment and involvement to some of the needs identified for future research and extension programs.

Existing Data—Do not overlook the existing local farmer's knowledge, sales and marketing patterns, uses of agricultural resources, and social behaviors within communities, as a means of involving farmers and others in the nine cells. By looking at existing data and using them, one is indirectly involving the farmers in the process.

Concluding Comments

Participation of farmers in extension and research is inevitable and effective. This chapter has provided evidence and arguments for that conclusion. More specific and crucial questions are therefore in need of discussion, which also have been discussed in this chapter. Those more specific involvement questions are related to both program functions and program phases. Farmers, extensionists, and researchers perform different roles and tasks in involvement because of their different knowledge and experiences, depending on what function and what phase of the program it in.

Within each cell, different techniques can be used to get people involved in the research and extension processes. Certain techniques involve farmers and others as individuals and others involve them in groups.

Several important points about involving others must override all of what has been said.

First, keep in mind your purposes and potential benefits for involvement as were stated early in this chapter. Select the time, the techniques, and settings according to what your purposes are for involvement—learning? planning? evaluating? sanctioning? being ethical?

Secondly, listen and observe. Listening and observing are not separate techniques. Listening and observing needs to happen in all of the above techniques and nine cells. No good comes of involving farmers physically and socially if their concerns are not heard or observed, and internalized, and then acted upon.

Thirdly, involve the people who are in the best position of acting and follow up. For instance, if women are doing certain agricultural tasks more than men, they must be involved. If leaders in a village are one to make sure that follow-up occurs, get them involved. If researchers believe their time is wasted by involvement processes, get them involved and demonstrate the power of total involvement by everyone to bring about rural development.

References

Ahmad, A. "Participation of Adult Learners in the Program Decision-Making Process: A Study of Felda Schemes in Malaysia." Unpublished doctoral dissertation, University of Wisconsin-Madison, 1987.

Bamburger, Michael. *The Role of Community Participation in Development Planning and Project Management,* published by the Economic Development Institute of the World Bank, EDI Policy Seminar Report, The World Bank, Washington, DC, 1986.

Cellman, L. *Listen to the People: Participant Observer Evaluation of Development Projects,* Washington, DC, The World Bank, 1987.

Compton, J.L (Ed.). "Integration of Research and Extension," in *The Transformation of International Agricultural Research and Development,* Lynne Rienner, Boulder and London, 1989, p. 123.

Crouch, B. R. "The Problem Census: Farmer-Centred Problem Identification", FAO, ILO, UNESCO: *Training for Agriculture and Rural Development,* 1983.

Delbecq, Andre, Andrew Van de Van, D. H. Gustafson. *Group Techniques for Program Planning: A Guide to Nominal Group and Delphi processes,* Glenview, IL, Scott, Foresman, 1975.

Dewey, J. "The Continuum of Ends/Means—Theory of Valuation," *International Encyclopedia of Unified Sciences,* Vol. 2, No. 4, Chicago, IL, University of Chicago Press, 1939.

Forest, Laverne B. *Situational Analysis: Module 4, Working With Our Publics,* North Carolina State University, Raleigh, NC, 1988.

Forest, Laverne B. "Community Leaders' Commitments to Programs: Do They Change During a Program? In *Adult Education,* Vol. XXVIII, No. 4, 1978, pp. 211-222.

Forest, Laverne B. "Adult Education Program Commitment: What Is It? What Causes It?" Paper presented at International League for Social Commitment in Adult Education, Gothenburg, Sweden, July, 1985.

Forest, Laverne B. "Judgments About Extension", *Shawano County Leaders,* No. 9, University of Wisconsin Extension, Madison, WI, 1978.

Forest, Laverne B. "Involvement in Extension Program Development, *Shawano County Leaders,* No. 6, University of Wisconsin Extension, Madison, WI, 1978.

Forest, Laverne B. "Establishing Goals at the Local Level", in *Examining Controversies in Adult Education,* editor B. W. Kreitlow and Associates, Jossey-Bass, San Francisco: CA, 1981, pp. 186-195.

Forest, L. B., and Mary Marshall, *Impact of Extension in Shawano County: I. Conclusions and Implications,* Madison, WI, Division of Program and Staff Development, University of Wisconsin, Extension, 1977.

Forest, L. B. "Present Commitments: Their Relationships to One's Acceptance of Community Change and Programs." *Adult Education,* Vol. 23, 1973, pp. 171-191.

Forest, L. B. *Criteria for Good Agricultural Extension Programs: Do they Transcend Culture?* Paper presented at American Evaluation Association, San Francisco, CA, October, 1989.

Osborn, A. F. *Applied Imagination: Principles and Procedures of Creative Problem-Solving,* 3rd Revised Edition, NY, Scribner, 1979.

Paul, Samuel. "Community Participation and Development Projects: The World Bank Experience, in *Readings and Community Participation,* Washington, DC, EDI, Chapter 2, 1987.

Sail, Rahim Mohammed, and Laverne B. Forest, "Experiences that Relate to Adult Education Programme", in *Land Development Digest,* Malaysia, Vol. 7, No. 2, 1985.

Whale, W. B. "Appraisal of a Process of Planning for Total Resource Development in a Wisconsin County," Unpublished doctoral dissertation, University of Wisconsin-Madison, 1966.

11

Some Thoughts on Farmers' Participation in Agricultural Research and Extension

S. B. Chattopadhyay

During the past three and half decades, significant development in agricultural production has taken place in India. This dynamic change has been possible through scientific planning, intensification of agriculture and adoption of scientific technology together with use of HYV seed, fertilizer technology, increased facilities of irrigation and development of infrastructure.

The pace of development has not spread uniformly in all parts of country, particularly Eastern India. The agricultural potential of Eastern India comprising West Bengal, Orissa, Bihar and East Uttar Pradesh is rated very high—rich soil, abundant manpower, adequate rainfall, facilities of greater utilization of water through both surface and groundwater irrigation. The technology of high production using varieties of crops with higher yield potential does not seem to have taken roots in the manner it has been expected to do.

Historical records reveal that this region was a prosperous and agriculturally advanced one in the country. Even after independence, the eastern region recorded the highest level of productivity (yield/ha), but as would be evident from the Table, it is gradually showing much less increase in productivity, as compared with the other regions of the country, particularly Northern and Southern regions which took full advantage of the new technology of intensive crop production.

Region	*Trends in Foodgrain Yield (kg/ha) Triennium Average*			
	1950-51	*1960-61*	*1970-71*	*1980-81*
Eastern	644	765	897	970
Western	390	524	551	649
Southern	554	731	897	1149
Northern	608	788	1150	1493

Comparatively lower level of productivity has been attributed to some extent due to high risk and heavy managerial responsibility upon the farm operators far beyond the capacity and capital resource of small and marginal farmers. Even at the currently available level of technology, the difference between the yield in the field of farmers under National Demonstration and average yield varies by a factor of 3 to 6.

Regarding utilization of modern technology, two factors have been recognised, namely, first, adaptability or suitability of a technology in a given locality wherefrom the concept of location specific technology has arisen and second, economic scale which has focussed attention to the development of intermediate technology suited to groups with limited resources or limited capacity to utilize resources and investment.

It is also necessary to take into account the cultural base. The technical know-how and infrastructure cannot make a headway in a locality until it finds compatibility with prevalent cultural dynamicity. To illustrate this point one may compare the operation and performance of deep tubewell in Haringhata P. S. in Nadia district with those of in Narayangarh P.S. in Midnapur district, West Bengal. In case of farmers of Nadia district, the farmers are owners of the land, they have a minimum level of education and agriculture is more remunerative, hence the utilization of ground water has been optimum, while in Narayangarh P.S., the utilization has not been upto expected level. One of the major causes of variation is cultural. The culture that prevails in Narayangarh P.S. is characterized by numerical domination of scheduled castes and tribes who own very little land, but carry on agricultural operations as share croppers or agricultural labourers. Technical know-how may be put in a capsule and this capsule may be used for exploitation by technocrat. But from a given socio-cultural frame of reference, that capsule cannot be sent to extension workers for better utilization of resource without consideration of a society and culture. This is because of the reason that the degree of variation is too high or level of acceptance is too low with regard to socio-cultural factors that determine the human behaviour. Hence question of participation of farmers in research and extension programme arises.

In this process we come across two sets of factors,

No farmer involvement (Research base)			*Farmer Involvement (Extension base)*	
Physical	*Biological*	*Support System*	*Resource*	*Social*
Water availability	Varieties	Extension		Tradition, goals
Temperature	Diseases	Credit	Land	
Sunshine conditions	Pests	input	Labour	
Soil conditions	Weeds	Price	Capital	
		Govt. Policy	Management	
Cropping pattern evolution		Development of cropping system		

Activities which are concerned without involvement of farmer lead to evolution of an ideal or suitable cropping pattern. Whether this cropping pattern will fit into the system of the farmers forming a new cropping system depends on the participation and interaction of farmers regarding facilities available for the purpose within a given socio-economic framework.

Crop intensification research has two facets. Its aim is to provide the requisite technology for achieving the goal, but it has also another responsibility, namely transfer of technology to user agencies in such a manner that can be easily and rapidly spread to farmers.

Intensive cropping pattern system depends heavily on management. Hence farmers' existing and potential resources as well as crop management capabilities must be appraised. Hence development of "small farm" technology should be evolved.

Evaluation studies were carried out by National Council of Applied Economic Research in 1980 in West Bengal on Indo German Fertilizer Educational Project of Hindusthan Fertilizer Corporation. Some of the pertinent facts that emerged out of the study are:

(a) Where assured irrigation is available, small and marginal farmers are more-efficient users of irrigation water than medium and big farmers.

(b) Even if there is inadequacy of irrigation water in the field of small and marginal farmers, cropping intensity is more.

(c) Small and marginal farmers are more interested in use of fertilizers and they use more complex fertilizers than medium and big farmers, though the level of consciousness is same amongst all levels of farmers.

(d) Small and marginal farmers are more efficient users of fertilizers in comparison to medium and big farmers.

(e) For farming operations, small and marginal farmers employ more labour, where as medium and big farmers use animals and machines.

(f) There are comparatively greater opportunities of work among small and marginal farmers than among medium and big farmers.

(g) Productivity is higher in the fields of small and marginal farmers than in those of medium and big farmers.

Thus contrary to our belief and expectations, there are qualitative differences in attitude and behaviour of small and marginal farmers and medium and big farmers, though the resources and financial resources are more in the latter group.

There must also be sufficient component technology in the form of pest management—choice of appropriate varieties, adaptation of crops to different soil and environment, soil fertility management etc. The component technologies thus evolved should be assembled into functional cropping system. This is a far more complex matter which depends on limitations of management abilities of farmers as well as physical and socio-economical environment. The findings should be tested under conditions where we expect them to fit. This will necessitate on-farm research in the field of farmers, where it will be possible to place evolved technology in relevant conditions to take advantage of skill of farmers in the process of learning and evaluation.

To carry on the above mentioned on-farm studies, two approaches are necessary—first, approach should be made to study the farmers' system, how the farmer thinks and how he functions in his own environment—physical and socio-economical. Second, the new technology is to be incorporated in the farmers system then through analysis and evaluation, it has to be ascertained how it fits into and affects overall farming system. The entire study can be fruitful, only when the farmer actively participates in the whole process.

These observations definitely point out that participation of farmers is an imperative requisite for successful adoption of any new technology or strategy connected with increase in productivity in agriculture. In planning research programme, felt needs of the farmers are percolated through normal extension channel to the organization on personnel connected with planning and implementation of research activities. So far programme and its execution are concerned, as they stand, an inductive method is followed by and large. Involvement of farmers in on farm research has not been upto desired level at least in some parts of the country. The net result has been either the programme has not resulted in any fruitful benefit or the programme which had the potentiality and would have been successful have met with failure. An example may be cited of failure of the programme of introduction of medium staple cotton in the coastal belt of West Bengal in the rabi season following main crop of paddy which met with failure because it was introduced almost straight forward without any adaptive research or on-farm research in the field of farmers to find out the feasibility and without developing component technology. Similar fate probably awaits sugarbeet in the same area, probably for the same reasons.

Another example may be cited of wheat crop in West Bengal. Following introduction of high yielding varieties of wheat in West Bengal, from an area of 40,000 - 50,000 hectares prior to introduction of HYV, the area shot up to 5.65 lac hectares in 1975-76 and the total production also rose from 35,000 tonnes in 1965-66 to 11.87 lac tonnes in 1975-76 and productivity from 700 kg to 2100 kg per hectare. But area productivity and total production gradually showed decline as would be evident from the following table. It may be worthwhile to point out that West Bengal is deficit in foodgrains as a result food-grains always fetch a good price in the market.

Wheat:	*1975-76*	*1976-77*	*1977-78*	*1978-79*	*1979-80*	*1980-81*	*1985-86*
Area	5.65	5.15	4.84	5.21	5.05	2.83	3.05
Total yield	11.87	10.51	10.36	9.98	7.68	4.73	5.77
Yield per ha	2100	2040	2140	1916	1518	1672	1894

(Area in lac hectares, total yield in lac tonnes and productivity kg/ha.)

This situation has arisen mainly due to lack of technological support which the farmers needed to keep up the cultivation of wheat crop and inadequate realisation of needs of the farmers in this context.

Hence it is felt that we have lab to land programme, later on probably renamed as TOT, a somewhat similar programme in the reverse direction should be organized—Land to Lab. Agencies should be set up with farmers at the Block, District and State level with representatives of farmers of different categories in which contact or coordinating officials will be present. Discussion on problems facing agricultural production can be held from the view point of farmers and they can be sorted out on priority basis and passed on to authorities for appropriate action. These organizations would by and large farmer dominated bodies.

In this context it may be pertinent to point out that demonstrations in the field of farmers which formed the basis of recommendations in connection with transfer of technology are not receiving due attention at least in some states under T&V Programme. These should not only be properly looked into, but also augmented to have better reaction between the farmers and technologists, as well as more effective participation of the farmers in research and extension programme.

12

Farmers-Researchers Participation and Collective Actions—A New Dimension in Technology Transfer for Agricultural and Rural Development

R. K. Samanta

Introduction

About three quarters of the world's population live in the developing world and of these three quarters—some 2800 million people—live in rural areas. They represent the majority of humanity, and their numbers are growing by about two per cent a year. Sustained chiefly by work on the land, they produce not only food for themselves, but much of the tropical food and raw materials consumed in developed countries. Their output is vital to national economies and world prosperity.

Yet rural areas also account for most of the world's hunger and misery. A recent study by the Food and Agriculture Organisation found that "even all optimistic assumptions, at least 700,000,000 rural people are now living in absolute poverty." The vast majority of the rural poor subsist through farming, dairying or fishing; they are peasant farmers, tenants and share croppers, rural artisans, fisherfolk and landless workers. Their basic characteristics are the same: they have limited access to natural resources, modern technology and services; they lack institutions which would sustain higher levels of productivity; and they are usually found living alongside a relatively prosperous privileged minority.

Rural Development, therefore, is a strategy designed to transform rural life by extending to the mass of the rural population the benefits of economic and social progress. It clearly endorses efforts to increase production and raise productivity in rural areas. But it stresses a fundamental principle: that the rural poor must share fully in this development process through equitable access to resources, inputs and services and participation in the design and implementation of development programmes.

For years, observers and critics of development efforts have been calling for greater participation by the intended beneficiaries in the design, implementation and evaluation of projects as it is necessary that without participation from the beneficiaries no development effort can achieve its desired objectives.

The participation of rural clientele in the development, implementation and evaluation of extension/rural development programmes has been a long standing, fundamental guiding principle. It is a principle which represents the very essence of the non-formally, popular nature of rural development work. And there has not been a wide degree of adherence to this principle throughout this country and some other parts of the world. It is fact and proved that there can be no development unless the urban and rural poor want to develop themselves. And they will highly be keen to participate in the process. They have notable hand in shaping or formulating. The present day world wide sweep of humanistic, democratic ideology has, therefore to be linked with modern technology and socio-economic concepts for serving the human needs and aspirations for a better life of a teeming millions. The agricultural/rural development, therefore, must combine with growth and equality. These rural masses are quite receptive to new ideas and they want to be informed, despite high rate of illiteracy in them. So, it is necessary to serve the participation of these rural gentry for stabilising the impact and gains of the improved technology for increasing production and productivity and to have a balanced and even economic growth.

The rural masses have always been exploited in the past, in every one, who approached them and thus they have developed a sense of aloofness due to suspicion. They feel that they are treated as third rate citizens. He hesitates to believe the extension man, as he knows very well, that advisor has no personal financial stakes, and in case of the failure due to his advice, the farmer is the looser.

Repetitions of such failures forces him to become conservative and to go in the way he has known. He is not dull, docile, unintelligent as has been thought by the urban intelligentia. Of course he is over sensitive and individualist. He makes his own decision. The package approach, amounting to Government proposing and farmer disposing, as if to a minor child, disregards the need of the farmers' conviction. He is fully grown up adult man. He has exercised his franchise so many times and has brought down many big personalities. It has to be realised that the Government can propose, pioneer, promote and sponsor schemes, but the performance has to come from the farmer alone. The performance cannot come without participation and meaningful participation cannot come without conviction and identification based on personal practical experience.

It is only when this new approach is followed that we can expect the people to come forward and willingly participate in rural development programmes. Their experience so far has been that the benefits of development projects emerging out of their participation have gone to the upper strata in the rural areas. Peoples' participation becomes meaningful only when such a common core of values is not only evolved but also communicated to all sections of the society through education of masses and folk media as well as through demonstrated adherence to these norms by the modernizing elite. Peoples' participation involves promotion of voluntary organizations and associations of persons with common interests.

Therefore, their participation from the very beginning of the planning of any rural development programme is essential for achieving its targets. The present paper will look briefly at the rationale for participation; farmers-researchers interface; farmers' participation & collective actions; and newer technology development, transfer and adoption for all round agricultural and socio-economic development of the clientele—a vast majority of rural masses spread throughout the world.

Why Participation?

Participation by intended beneficiaries in the agricultural development programmes may yield the following advantages and positive outcome.

1. During the design phase of a programme, intended beneficiaries can provide cultural, ecological, economic and technical

indigenous knowledge. Designers or planners can avoid costly mistakes as they and the beneficiaries determine which techniques and programme objectives are most appropriate for the community.

2. Programme participants can mobilize the local resources, in the form of cash, labour, materials, managerial talent and political support which are critical to programme success. Gow and Vasant (1983) argue that voluntary commitment of their resources is a necessary condition for breaking patterns of development paternalism which reinforce the people's passivity and dependency. Farmers and peasants are more likely to contribute resources to a project they have helped select and design than to one chosen by an outsider.
3. The fields and the livestock pens of the farmers are the best testing grounds for proposed agricultural innovations. Farming Systems Researchers have concluded that onfarm trials are more appropriate to local needs, faster and less expensive when farmers participate in all stages, including experimental design, of those trials (Ashby, 1986).
4. Programmes involving participation are more likely to be sustained after outside funding is reduced or withdrawn, or after the outside technicians or change agents leave the scene. The motivation, training and financial arrangements necessary to maintain programme activities can best be developed when the beneficiaries have been involved in the programme from the beginning.
5. When community members participate in the programme's implementation phase as para professionals, they extend the benefits to a greater number of people than scarce professionals could reach. Some new ideas are more readily accepted when introduced or demonstrated by trusted Peers.
6. Participation by the poorer elements of society may prevent the "hijacking" of programme benefits by wealthier members of the community.

Farmers-Researchers Interface

Indigenous research that addresses directly the problems and needs of small farmers is at the core of agricultural development. The history of research and development for increasing human productivity on the marginal subsistence farm has pointed to systems approach. The small farmer is a part of the larger system that surrounds his day-to-day activities. These activities include household concerns that direct priorities and decisions the farmer incorporates in his farming system. Systems logic is needed to solve systems problems. Logically, the solutions will be systems solution. Much of previous experimentation have been linked to conventional methods of research practice on experimental stand models. This type of research has been very successful in developing new technology. But this technology has seldom been adopted. A great deal of money has been wasted on R & D projects throughout undeveloped, under-developed and third world countries. That produces nothing useful for the small farmers.

The first step in designing and experimenting for research is determining the nature of the problem. What is the issue then? The same question must be asked of the R & D process in developing countries. This issue is of resources poor farmer. Perhaps, the description of resource poor, is a distorted terminology. The farmer as an individual or a member of a group or family is rich in resources. But greatest resource of a region is human. Not merely the physically potential of motivated mass is a resource but the stored knowledge and idea sources of the human mind.

The assumption is who benefits from research and development. Agricultural Research has served the resource rich farmer. They are rich in resources of land, labour and capital and they are provided with a number of juicy options for increasing production. On the other hand, subsistence farmer is on the margin and can not buy into the production benefits supplied by current agricultural research. The small farmer creates a safety net in survival by diversing the farming system. The integration of crops, animals and household labour provides them a consistent supply of food during the uncertain cycles of rural life. These must be known by the researchers before they try to develop the technology for the farmers.

The small farmer uses systems group strategy to solve their problem. The strategy used by much of scientific research for small farmers has focused on the individual farmer and individual group. This lack of appropriateness can partly be attributed to the most basic assumption made by scientific research. The assumption is that something of value will be given to another. Therefore, the role of the researchers and the farmers are required to be changed. This might be called reversal.

The systematic changes needed in scientific re-orientation began with this rural reversal. The researcher must first learn from the farmer.

The researcher must understand and reconcile the consistent facets between his knowledge and that of the farmer. The transfer of technology must pass from farmer to scientist on the farm and in the farmers, community. The scientist can not learn from the farmer on the research station or in laboratories. They must meet face to face in the farmers' environment. Because of the more cultural gap between them there have been many cases of farmers saying 'yes' but still meaning 'no'. The scientist must make the reversal and meet the farmer because the farmer is bound by cultural traditions that prevent him from open up to a scientist. Therefore, the issue is not only one of transfer of appropriate technology to farmers but the transfer of appropriate scientists.

Farmers' Participation and Collective Actions

First we must know what their real priorities are, how they weigh what is desirable against what is possible, what risks they are prepared to take.

Once development is approached from the bottom up, the distortions in the conventional approach of development from the top down quickly become evident. Factors which seem marginal from an overall perspective assume great importance: in the case of the peasants, for example, it is their history as producers, their achievements and failures, the logic and rationale of their decisions, and hence their real interests and objectives as individuals, families, communities and social group.

Development from the bottom up is thus a process which starts on a human scale. This is reflected in the geographical dimension of a programme: the rural micro-region where all the peasants' real contacts occur, or the district in which the shanty-town dwellers spend their lives.

Also important is the social setting in which a programme is implemented: the local community, neighbours who have daily contacts with each other. Nothing could be further from the world of abstract categories. Every peasant or shanty-town dweller has a face and a name, is something more than a statistic.

It is reflected, too, in the integrated nature of the development programmes, in which agricultural production is one of the problems but not the only one, and is closely linked to education, health, housing, culture, recreation, local social relations, etc.

This search for the human scale, this concern with development from the bottom is also expressed in the selection of the tools to be used: the precedence given to local resources, the development of groups' own capacities, the recovery of the valid elements in traditional technologies, the search for appropriate technologies based on local resources—technologies that can be managed, understood, improved and transmitted by the basic groups themselves.

In this approach, development at the local level, on a community scale, is also seen as an apprenticeship for tackling problems at higher levels. It will be easier, for example, for peasants to understand the obstacles to their development in the society as a whole if they discover them through analyzing the difficulties that cannot be overcome at the local level. In the same way the importance of their organization as a social class will not be just a theoretical abstraction if they have first hand positive experience within their local community.

Since development is a process in which there are material objectives to be achieved, and since these should benefit first and foremost the largest sector of society, the populace—the peasants and farmers, believe that if development is to be real and lasting, and to grow, these material objectives must be attained by getting the peasants and farmers to act as the main social protagonist of the process.

The peasants and farmers must assume responsibility for meeting its own requirements, produce the goods it needs and thus develop itself by becoming increasingly capable of tackling bigger problems.

The concept of the people as protagonists therefore contrasts with that of the people as the subject of an action conceived, decided upon and carried out by others. In other words, development of the people can only be achieved if they attain sufficient autonomy and participation in society.

This would mean an enrichment of society, in the sense of an organized social structure, and enable it to act as a counterpart to the State: a counterpart which, depending on how it reacts to its own problems, will be able to modify the State if the latter does not respond to its interests, or to control it and make an organized contribution to overall development.

Farmers' Participation for Appropriate Technology Development, Transfer and Adoption

Now the basic question arises, what does the populace need in order to become the protagonist of its own development?

First and foremost, it needs to exist as a people. It cannot be spoken of as a social protagonist if it consists solely of separate individuals, with no solidarity or cooperation between them.

A people exist when they have managed to develop their own organizations, from the most elementary and circumstantial, constructed on the basis of very concrete and immediate problems, to those at higher levels, able to tackle society's overall problems. They must have an organizational network, a social fabric that is the expression of their existence as a people and which at the same time guarantees their strength, their ability to act effectively within society and to be listened to.

To build themselves into a social protagonist the people also need to know where they are going. They need to know what they are as a social sector, who are in the same conditions and who are not, who have historically been the cause of their situation and hence who have interests opposed to theirs. They need to know what changes have to be made in society, what society they aspire to, what path will lead to it, and what means they can use to achieve it. In a word, they cannot be protagonists unless they develop a social consciousness which, like organization, arises and grows from small experiments, reflection on action carried out, successes and failures.

In this reflection a people's history provides valuable elements for understanding the present and illuminating the future. But social consciousness is not just understanding, it is also commitment: commitment to the development of one's group, one's social class. That is why social consciousness, like organization, develops and expresses itself through group action.

The development of organization and social consciousness do not seem to be enough. However, a subject people is not a social protagonist nor is a dependent people; since dependency is another form of subjection. A people unable to do anything positive about its needs, particularly the most basic ones, is a subject people.

The central factor in what we have called development from the bottom up is, to a large extent, a growing initiative at the grassroots, which means that the process starts from there.

The poor will hardly be able to develop their initiative if they are unable to obtain significant degrees of autonomy vis-a-vis their most basic needs: an autonomy that will enable them, for example, to negotiate with other social groups; an autonomy that will signify some freedom, some independence from the state, and enable them to have their own personality, capacity to control, capacity to contribute.

Achievement of this autonomy implies development of the people's own capacities, in the sense of both material and human resources. A people who start from what they are and what they have in order to develop can become the protagonist of the development process.

One question remains. *Can* the people develop the ability to meet their requirements by their own efforts?

The reply to this question depends on the way the tasks are tackled, and the means selected to resolve the problems—i.e., on the technological option chosen.

Conclusion

Experience has shown the value of technologies that emphasize local resources rather than contributions from outside; technologies that are based mainly on human labour, which is much more abundant than capital among the poor; technologies that start from the experience accumulated by the people and reinforce their culture, particularly their values, objectives and interests; technologies that are understandable and manageable and hence permit and stimulate creativity. In transferring them, therefore, the accent should be put on understanding the principles and criteria on which they are based, rather than on readymade formulas and recipes.

Such a technology can generate greater autonomy and independence in the basic groups. The technological option is therefore not neutral, since it may or may not beget freedom.

For example, if the technology to be used takes advantage of local resources, the group will have more resources than if a technology that does not make use of them were chosen. This difference may result in a group considered marginal becoming a sector able to contribute to the development of the country.

We said at the beginning that the point of departure for this approach to development from the bottom up must be the people's own will to develop. We think there are two things that can kindle this will. The first is that development should seem desirable to the people. The second is that it should seem possible. To be desirable, the objectives proposed must be related to the people's interests and achieved in a way that does not contradict their logic, their way of reasoning. Similarly, the technology proposed will to a large extent determine whether the group sees it is possible to start its development process itself.

The technology may also affect organization and social consciousness. A technology manageable by the basic groups can be of great help in developing the basic elements in organization: responsibility, planning, group discussion, joint decisions, control of the tasks done by the same group management development, the emergence and training of new managers, and many others.

To sum up, an appropriate technology makes it possible for social awareness to be, not a phenomenon detached from the daily routine, an abstract idealization, but something that arises from, and is developed by day-to-day experience.

Understood in this way, the technological option and the way it is used becomes a key element in our programmes, since it can make it easier or harder for the people to become the protagonists of their own development, making an approach to development from the grassroots up feasible or, on the contrary, Utopian.

References

Ashby J. 1986. "Methodology for the participation of small Farmers in the design of On-Farm Trials", *Agricultural Administration,* 22: 1-19.

Chambers, R. 1973. *Managing Rural Development: Ideas and Experiences for East Africa,* Uppsala: Scandinavian Institute of African Studies.

Gow, D. and Vasant, J. 1983. "Beyond the Rhetoric of Rural Development Participation: How Can It Be Done?", *World Development,* 11: 5, 427-446.

13

Extension of Farming Systems Research Technology in Bangladesh

Abdul Halim and Md. Shahidul Islam

The inception of green revolution has led to adopt technologies requiring substantial use of irrigation and agro-chemicals. In recent years, it has become clear that a large number of resource poor farmers gained little from these technologies. One may still contradict with this view especially under Bangladesh condition as it was found in a survey that farmers operating less than one hectare of land allocated 52 per cent of their rice land to modern varieties compared to 45 per cent for those with one to two hectare and 42 per cent for those with over two hectare of land (Hossain, 1987). In spite of the higher adoption of modern varieties by the small farmers, they receive little benefit since they are to pay higher charges for all inputs and receive lower price for output. In addition, they can hardly apply inputs timely and as a result receive low yield and production. In India, there is a widespread apprehension that the diffusion of new technology would contribute to worsening income inequality.

The technology developed by the farming systems research (FSR) team in Bangladesh has got a basic difference with that developed by the commodity or discipline oriented on-station researchers. The FSR technology is problem and system oriented. The FSR is a farmer participatory research carried out at the farmers' own environment by an inter disciplinary team in holistic approach. The technology developed by FSR scientists is thus based on the farmers physical, biological, social and economic environments which is particularly suitable for a specific group of farmers for whom the technology is developed. The FSR technology may, therefore, be defined as a technology immediately needed by a group of farmers called a client group and may/may not be suitable for all others. The methodology of extension of FSR technology is thus most likely be different from that of a conventional technology although a considerable similarity in extension of these technologies is not unlikely.

Why to set up a separate methodology for extension of FSR technology in spite of the existing technology extension system needs to be understood first. Actually, extension starts from the inception of a particular technology and continues upto the final adoption. The course of development of a particular technology has an impact on the extension of that technology. The course of development of an FSR technology from its inception to maturation is completely different from that of a conventional technology. It thus goes without saying that there should be a clear cut methodology for extension of FSR technology. This chapter will deal with the principles and procedures for extension of FSR technology and with a comparative study with the existing extension system of technologies developed through commodity/ discipline oriented research.

Extension of a particular technology is as simple as complex. In this connection, the information about existing technology adoption systems followed both by governmental and non-governmental organizations need to be reviewed which will help develop a viable methodology for extension of FSR technology.

Existing System of Technology Extension

Among different government organizations involved in extension of agricultural technologies, the names of Department of Agriculture Extension (DAE), Directorate of Livestock Services (DLS), and Directorate of Fisheries (DF) are worthmentioning. However, only the DAE has village level workers to work directly with the farmers. In reality, this is the only organization engaged purely in extension of agricultural technologies. The organogram of the extension wing of this organization is provided in Figure 1.

The approach of transferring technology is a typical Training and Visit (T & V) system where the impact points for extension are determined by the Subject Matter Specialists (SMS) and Scientists of different institutes through a workshop at the district level which receive a final approval from the District Technical Committee (DTC).

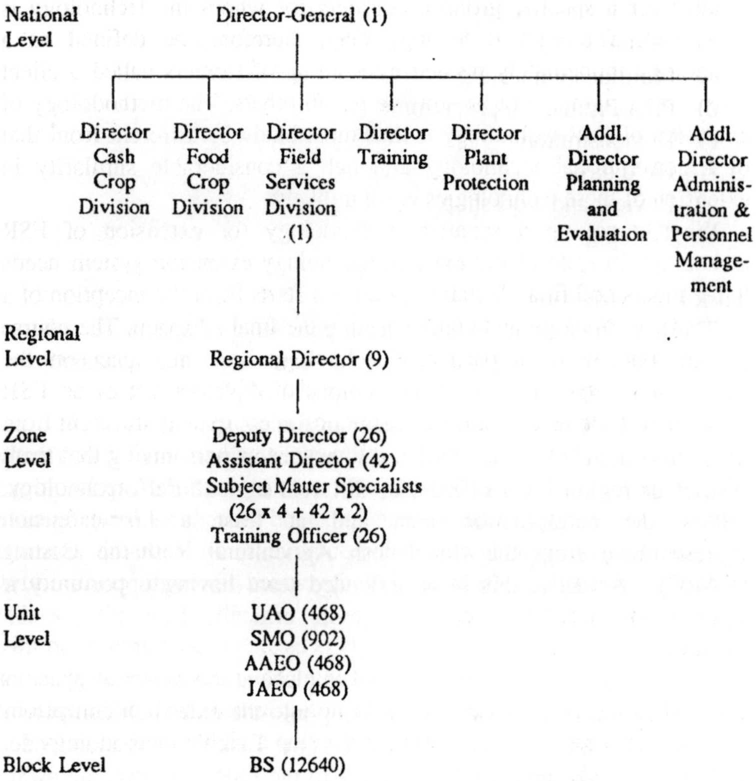

Fig. 1: Organizational set up of extension wing of DAE.

Upon approval, the SMS train the Subject Matter Officers (SMO) for onward transmission to the Block Supervisors (BS) and ultimately to the farmers. This approach may be called the "top down" approach and suffers the set back of the lack of farmers' participation for which the real problems of the farmers would not reach up to the researchers. The great assumption in this system is that the contact farmers who are usually the rural elites or innovators have the capability of adopting and/or disseminating all those technology tablets/developed by research institutes. Since the technology is mostly developed without considering the farmers resources, the extension workers do not consider the ability of the farmers. This gap of the existing extension methods being practiced in Bangladesh through T & V system may be taken care to a great extent by the FSR system. The important elements which are considered as extension component of FSR are

a) Task Force meeting
b) On-farm trial
c) Multilocation testing
d) Pilot Production programme
e) Demonstration
f) Field day
g) Review and workshop
h) Training

Task Force Meeting

The task force of an FSR is a team consisting of specialists relating to all the developmental research, extension and management organizations/agencies. The team composed of all the members of the FSR team, FSR practitioners of other institutes, specialists of different areas not included in the FSR team, extension personnel specifically district or region level officers of different agricultural organizations within the extrapolation area and at least a knowledgeable representative from the Bangladesh Agricultural Research Council (BARC). Actually, this is an extended team having opportunity to examine the research programme more critically from all possible angles.

The meeting of the Task Force takes place at the planning phase of FSR. Thus the inclusion of this meeting into the extension component of FSR technology needs justification. As the Task Force is an extended FSR team, the members not belonging to the FSR team become aware about the farm environment and farmers problems, ideas and opinions which ultimately help them to plan specific research related to farmers problems. In addition to the information about farmers resources, the previous experiences of FSR are also considered for planning future research and as a result the members of the Task Force become well aware about the achievements of the previous years' research. Further, some of the members are also the members of the DTC, and the idea of Task Force meeting is automatically transmitted to the members of the DTC through their monthly regular meeting. In spite of these, it is more a component of research than that of extension. However, this definitely has an impact on extension of technology through different channels and as such is not illogical to be included in the extension component.

On-farm Trial

The FSR activities are carried out at the farmers fields or homestead areas as per programme after receiving approval of the Task Force. The research activities are managed by the farmers themselves as per instruction of the concerned FSR personnel. This helps the farmers to learn the technology by doing and by seeing. Although, researchers evaluate a technology in a prescribed manner, the farmers evaluate that in their own way, and if the farmers find the technology suitable, they go directly for adoption without concerning the research personnel and in some cases avoiding the researchers advice not to adopt. Actually, the farmers are more dynamic than the researchers especially in their farm environment. Whatever they find any technology suitable, they go for adoption without thinking the ultimate consequences. If the technology is really sustainable under the farm environment, it would be automatically diffused to other farmers. This tendency of the farmers has begun since the time immemorial and will be continuing for unforeseeable future. Most of the traditional technologies have been developed and diffused in this way. Therefore, this spontaneous diffusion of technologies should not be excluded from the extension component of FSR technology.

Multilocation Testing

Multilocation testing is the testing of a particular technology in different locations within the extrapolation area of the FSR site whose agro-ecological conditions are more or less similar. The same practice is done for the discipline-oriented or commodity oriented applied research. The basic difference between the multilocation testing (MLT) of a discipline oriented technology and an FSR technology is such that the former is tested under different locations of variable soil and environment irrespective of socio-economic condition of the farmers. The FSR technology is tested, on the other hand, within the same clientele or farmer group under variable soil and environmental condition. The performance of an FSR technology is measured considering the whole farm as a system as it has to be adopted into the existing farming systems of the group.

The multilocation testing of an FSR technology is needed for verification under different conditions and for further requirement of improvement if necessary. The advantage of multilocation testing of a particular technology is thus two fold, examination of performance of technology and suggestion to the FSR personnel about the direction of modification or improvement of the technology. The latter is the feed-back for the FSR. Upon recommendation by the MLT trial, the technology should go for a pilot production programme or directly to the production programme depending upon the nature and extent of risk involvement of the technology. If the technology is less risky, the extension personnel may recommend that for direct production programme. On the other hand, if there is any possibility of economic loss due to adoption of the technology, a pilot production programme should follow the MLT trials for about two years and then a production programme should be initiated.

The management of MLT trials needs be similar to the farmers' existing management condition. The extension personnel at the Upazila level is generally responsible. The Upazila Agriculture Officer (UAO) and the SMO need be informed well ahead of time so that they can arrange everything from selection of farmer and plot to the execution of the trials. They should be well trained about the execution and collection of data. All the materials needed should also be supplied in proper time.

Since the Site Co-ordinators of FSR are usually the members of the DTC, they can easily inform the members of the DTC about the MLT trials so that all the personnel involved in DTC become well aware about the technology. This needs to extend the DTC with the inclusion of DLO and DFO since FSR activities are not restricted to crop only. At the Upazila level there should also be a Co-ordination Committee for Technology Transfer (CCTT) consisting of UAO, SMO, ULO and UFO. This will be particularly helpful to test and transfer an FSR technology as it will provide a multidisciplinary committee which will be able to identify the farmers existing situation, need and problem in all directions.

The three principles of FSR namely, interdisciplinarity, holism and farmer participation should also be taken into consideration during MLT trials of FSR technologies. This is possible by forming interdisciplinary team at the Upazila level as discussed earlier, selecting participating farmers within the recommendation domain of the technology and providing opportunities to the farmers to play an active role in all the stages of implementation of MLT trials from planning to final evaluation.

The question may arise how MLT trials help dissemination of an FSR technology to the farmers. An easy answer to this question is the active participation of the farmers who learn the technology by doing and their neighbours learn by seeing. Moreover, the extension personnel receive information about the performance of the technology by themselves and they can include this in the impact point for the farmers whenever they think appropriate. Actually, this is the first step of extension of a particular technology and may be considered as the last step of research.

Pilot Production Programme

Pilot production programme is the production programme of a particular technology in a limited scale. This may be due to shortage of experimental materials, farmers' ignorance about the technology and risk in adoption. This can be considered as the second phase of extension and just one step before the adoption as the following step is the production programme. The pilot production programme stage is not essential for all the FSR technologies. This step may be skipped for some technologies which are relatively less risky and may easily be adopted by the farmers without much training.

The pilot production programme is implemented by the extension organization. In Bangladesh, the responsibility of carrying out a pilot production programme lies entirely on the DAE. The FSR technology, if considered mature enough after MLT trials, may go directly into the pilot production programme. The programme should be designed at the DTC meeting and be implemented at the Upazila level by the Upazila CCTT for testing and adopting the technology. Although this is purely an extension programme, the FSR scientists have got something to do especially at the planning and at the evaluation phase. The FSR scientists should monitor the programme throughout the course of the study so that they can identify the effect of this technology at the community level. They should also identify the requirement of further modification of the technology or the suitable method of extension of the technology. The evaluation results should be discussed in the DTC meeting to find out the real problems and to materialize the benefit of the technology. The contact farmers or the farmers involved in the pilot production programme need also to be interviewed for their opinion regarding the technology. Whether the technology is liked only by the farmers within the recommendation domain or by some other farmers need to be identified. If the latter is found to like the technology, the reasons behind this need to be found out which is not possible by a single researcher. This needs a multi-disciplinary team of researchers to identify the actual farmers who would be included in the recommendation domain.

Demonstration

Demonstration is an important step of extension of an FSR technology. It is likely that the farmers prefer learning by doing rather than by hearing or seeing. In this case some farmers get the opportunity to learn by doing. For others, there is an opportunity to learn by seeing. In extension teaching there are two methods of demonstration (1) result demonstration and (2) method demonstration. Although both these methods are considered equally important in extension of a particular technology, the extension of FSR technology needs the former more than the latter as the FSR technologies are mostly adaptive and are not completely new to the farmers.

The responsibility of management of the demonstration programme lies entirely with the DAE but the required inputs should be supplied by the FSR organization. The responsibility of the FSR organization is to monitor the demonstration units so that there is no mistake in technology adoption. The FSR personnel may also learn the reactions of the farmers regarding the technology. The demonstration plots should be distributed throughout the Upazila so that it covers all the representative environments (physical, biological, social and economic) of the Upazila depending upon the materials available.

Field Day

Field days or farmers' days are a part and parcel of extension programme of a particular technology. The methods of extension discussed earlier are not capable of involving mass people of the locality for adoption. So, a farmers' day or a field day is appropriate for disseminating the technology from the limited number of participating farmers to the general farmers.

The extension personnel are responsible for organizing the field day in presence of the research personnel especially the FSR personnel. In this mass communication meeting, the farmers may ask some more information about the technology which the researchers may have to answer. This meeting should preferably be a two way participating meeting instead of a one-way meeting usually done in the case of 'transfer of technology' approach. Since the FSR approach is a farmer participatory research, the farmers should be involved in every stage of FSR and extension of the technology. This does not need to change any physical set up of different extension organization except the formation of CCTT at Upazila level and the extension of DTC to the DLO and DFO.

Review and Workshops

Reviews and workshops are essential parts of FSR activities since these are important to the researchers as well as to the extension personnel. The review or workshop about FSR activities is done by the FSR personnel but is participated by both FSR and extension personnel and other agricultural research personnel of the region. This may be done to review a particular technology/several technologies tested or developed in a particular year/season. In the case of a particular technology, the FSR personnel and the extension personnel involved in developing the technology share their ideas and opinions among them and communicate these to the other researchers and extension personnel. In case of reviewing the whole activities of FSR and extension in a particular year/season, the aim is to discuss the advantages, disadvantages, problems and potentials of each technology with an ultimate objective to find out the promising FSR technologies.

In the extension-research review, results of FSR and MLT trials are discussed and if there is any technology found suitable after MLT may directly go for adoption, demonstration or pilot production programme depending upon the nature of technology. Since reviews and workshops are held at the region level, the DTC of each district within the region should include the important and useful points of the review related to the district in the discussion. The DTC is the authority to provide decision of what is to be done about a particular technology.

Training

Training is vitally important for transferring FSR technology. The extension personnel should be trained properly regarding the technology and situation of the farmers most likely to adopt the same. They should also be trained about the nature of development and adoption of FSR technologies which are usually different from the technologies developed by discipline or commodity oriented researchers. The 'transfer of technology' model is usually not fit for extension of FSR technology which needs an alternative approach may be called 'farmer first and last' or 'farmer back to farmer' approach.

Suggested Model

The suggested model for extension of FSR technology has been schematically shown in Figure 2. The model describes that the research findings come from the national research institutes through seminars, workshops, reviews and published materials are used by the FSR team

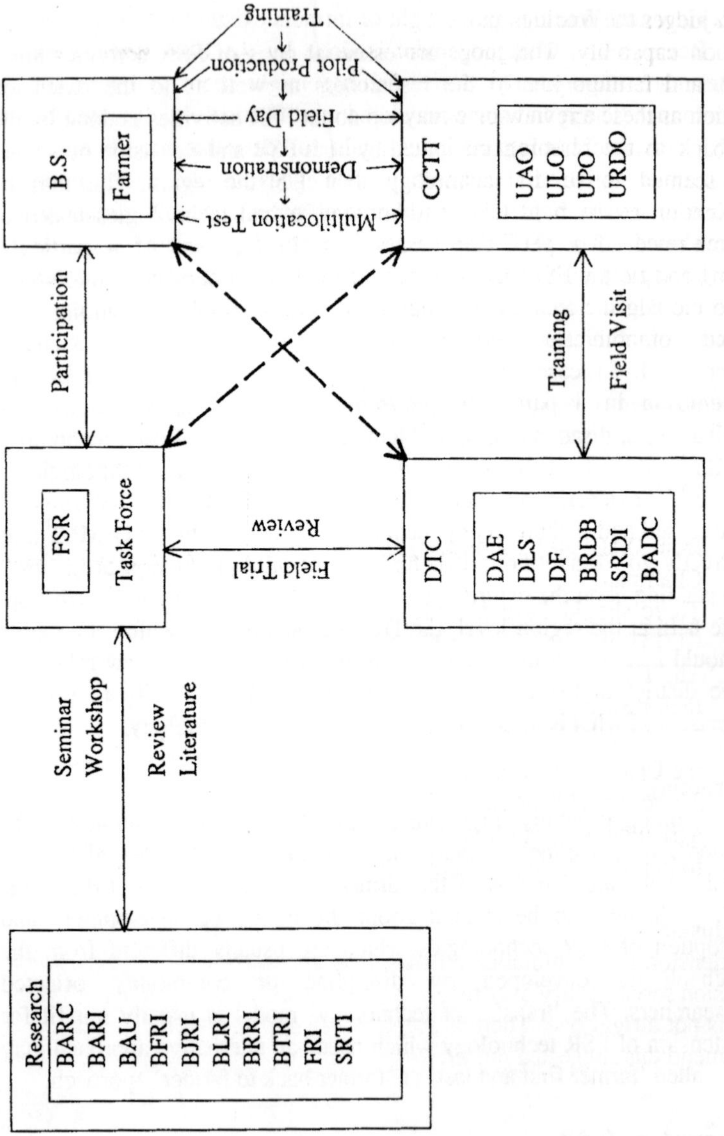

Fig. 2 : Schematic diagram of FSR technology extension system

which judges the findings in the light of farmers existing resources and adoption capability. The judgement is done by the FSR team, Task Force, and farmers jointly. Some findings might seem to be fit for adoption as these are and some may need modification which would be sent back to the research institutes again for desirable modification. This seemed to be fit technology then goes to the DTC upon examination at the field level and receives acceptance by extension-research review. The DTC then passes the technology to the CCTT via training and field visit of the members of the CCTT. The CCTT passes this to the Block Supervisors through different media depending upon the extent of motivation needed by the technology.

Elements Need to be Incorporated in the Suggested Model

1. The DTC should be extended to include DLO and DFO.
2. There should be a Co-ordination Committee for Technology Transfer (CCTT) at the Upazila level consisting of personnel from DAE, DLS, and DF. The senior most member may be the Chairman.
3. There should have a close linkage between the Chairman of CCTTs and the FSR Site Co-ordinators.
4. The FSR Site Co-ordinators should be the regular member of their respective DTC.
5. The Chairman of CCTT should have a regular discussion with the Upazila Chairman to obtain necessary logistic support.
6. There should be a memorandum of understanding among agricultural research institutes, universities, DAE, DLS, and DF to implement the programme at the field level.

Conclusion

Extension is a continuous process. The approaches and models for extension services vary over time and space. So, the suggested model of FSR is not a rigid one. There is ample scope and possibility to modify and amend this suggested model.

References

Box, L. 1987. Experimenting Cultivators: A Methodology for Adoptive Agricultural Research, ODI Discussion Paper 23. Overseas Development Institute, Regent's Park, London.

Chamber, R. and B.P. Ghildyal, 1985. Agricultural Research for resource—poor farmers: The farmer-first-and-last model. Agricultural Administration 2011-30.

CSRDP. 1986. Annual Report 1984-'85. Cropping Systems Research and Development Programme, Bangladesh Agricultural University, Mymensingh, Bangladesh.

Farrington, J. and A. Martin. 1987. Farmer Participatory Research: A Review of Concepts and Practices. ODI Discussion Paper 19. Overseas Development Institute, Regent's Park, London.

FSRDP, 1987a. Research Activities 1985-'86, Kazirshimla Site, Farming Systems Research and Development Programme, Bangladesh Agricultural University, Mymensingh, Bangladesh.

FSRDP, 1987b. Research Activities 1985-'86 Naogaon Site, Farming Systems Research and Development Programme, Bangladesh Agricultural University, Mymensingh, Bangladesh.

Halim, A., M.H. Ali and M. Islam. 1986. "A Study to determine the effect of T&V extension system on contact and non-contact farmers", Bangladesh Journal of Extension Education, 1 (1): 1-8.

Halim, A. and M. H. Ali. 1986. "The effect of on-farm research and extension on farm families in Bangladesh." Paper presented in the *Conference on Gender Issues in Farming Systems Research and Extension* held at Gainesville, University of Florida, USA during February 26 to March 1, 1986.

Hossain, M. 1987. Green Revolution in Bangladesh: Its Nature and Impact on Income Distribution, BIDS Working paper No. 4, Bangladesh Institute of Development Studies, Dhaka, Bangladesh.

Hossain, S.M.A., M. Sattar, J.U. Ahmed, M. Salim, M.S. Islam and M.U. Salam. 1987. Cropping systems research and farmers innovativeness in a farming community of Bangladesh. Paper presented at the IDS Workshop on Farmers and Agricultural Research: Complementary Methods, Institute of Development Studies, University of Sussex, Brighton, U.K. during 26-31 July, 1987.

Islam, M.S. 1987. Extension organization in Bangladesh. Bangladesh Journal of Extension Education Special Seminar Issue 2: 179-196.

Rhoades, R. and R. Prooth, 1982. Farmer-back-to-Farmer: A model for generating acceptable agricultural technology, Agricultural Administration 11(2): 127-137.

Abbreviations

AAEO	Assistant Agriculture Extension Officer
BADC	Bangladesh Agricultural Development Corporation
BARC	Bangladesh Agricultural Research Council
BARI	Bangladesh Agricultural Research Institute
BAU	Bangladesh Agricultural University
BFRI	Bangladesh Forest Research Institute
BJRI	Bangladesh Jute Research Institute
BLRI	Bangladesh Livestock Research Institute
BRDB	Bangladesh Rural Development Board
BRRI	Bangladesh Rice Research Institute
BS	Block Supervisor
BTRI	Bangladesh Tea Research Institute
CCTT	Coordination Committee for Technology Transfer
DAE	Department of Agriculture Extension
DF	Directorate of Fisheries
DFO	District Fisheries Officer
DLO	District Livestock Officer
DLS	Directorate of Livestock Services
DTC	District Technical Committee
FRI	Fisheries Research Institute
FSR	Farming Systems Research
JAEO	Junior Agriculture Extension Officer
MLT	Multilocation Testing
SMO	Subject Matter Officer
SMS	Subject Matter Specialist
SRDI	Soil Resource Development Institute
SRTI	Sugarcane Research and Training Institute
T & V	Training and Visit
UAO	Upazila Agriculture Officer
UFO	Upazila Fisheries Officer
ULO	Upazila Livestock Officer
UPO	Upazila Project Officer
URDO	Upazila Rural Development Officer

Abbreviations

AAEO Assistant Agriculture Extension Officer
BADC Bangladesh Agricultural Development Corporation
BARC Bangladesh Agricultural Research Council
BARI Bangladesh Agricultural Research Institute
BAU Bangladesh Agricultural University
[illegible] Bangladesh [illegible] Institute
[illegible] Bangladesh [illegible] Research Institute
[illegible] [illegible] Research Institute
BRDB Bangladesh Rural Development Board
BRRI Bangladesh Rice Research Institute
[illegible] Research
[illegible] [illegible] Research [illegible]
[illegible] [illegible] Technology Transfer
DAE Department of Agricultural Extension
[illegible] Department of Fisheries
[illegible] District Fisheries Officer
[illegible] District Livestock Officer
DLS Directorate of Livestock Services
DTC District Technical Committee
[illegible] Farming Systems Research
[illegible] Farming Systems Research [illegible]
[illegible] [illegible] Agriculture Extension Officer
[illegible] [illegible]
SMO Subject Matter Officer
SMS Subject Matter Specialist
SRDI Soil Resource Development Institute
[illegible] [illegible] Research and Training Institute
[illegible] [illegible] Village
UAO Upazila Agriculture Officer
[illegible] Upazila [illegible] Officer
[illegible] Upazila [illegible] Officer
UPO Upazila [illegible] Office
[illegible] Upazila [illegible] Officer

14

Change Agent Constraints in the Transfer of Agricultural Technology

G.L. Ray and P. Chatterjee

Introduction

A system of rules covering the rights and duties of employees and a system of procedures for dealing with work situations are two important dimensions of bureaucratic organisations (Bennis, 1966). However good the rules and procedures might be at the outset, these may require changes to improve organisation efficiency. A scientific method of making changes in the management of organisations would be to study the constraints perceived by its functionaries.

Transfer of technology is extremely important for modernising agriculture. In this context, a study has been made to find out the constraints perceived by the change agents in the transfer of agricultural technology to the farmers, so that the knowledge could be utilised in making appropriate changes in the management of organisations for increasing their efficiency.

In the present study, the change agent is a professional who occupies a crucial position in influencing the innovation-decision process of the farmers, who are the ultimate users of technology. Change agents provide the linkage between a change agency and the client system. Client system refers to the farming community who participated in the technology transfer programme, known as the Lab to Land Programme.

Methodology

Four organisations belonging to the Indian Council of Agricultural Research (ICAR) system and four organisations belonging to non-ICAR system, which participated in the second phase Lab to Land Programme of the ICAR in West Bengal from June, 1982 to May 1984 were selected. The Lab to Land Programme is essentially a technology transfer programme for which no separate staff was provided by the ICAR.

The organisations under the ICAR system were Bidhan Chandra Krishi Viswavidyalaya, Nadia; Palli Siksha Bhavana, Visva-Bharati, Birbhum; Jute Agricultural Research Institute, Barrackpore, North 24-Parganas; and Krishi Vigyan Kendra, Kakdwip, South 24-Parganas under the Central Institute of Brackishwater, Aquaculture. The non-ICAR system organisations, selected for the study were voluntary in nature. They were, Ramkrishna Ashram Krishi Vigyan Kendra, Nimpith, South 24-Parganas; Seva Bharati Krishi Vigyan Kendra, Kapgari, Midnapore; Ram Krishna Mission Lokasiksha Parishad, Narendrapur, South 24-Parganas; and Mandra Unnayan Samsad, Hoogly.

The total number of respondents (change agents) were 40, of which 28 were from ICAR system organisations and 12 from non-ICAR system organisations, comprising 16 scientists and 24 technologists.

The data were collected during August, 1985 to January 1986.

Dependent Variable

Attitude of the change agents towards transfer of technology was considered as the dependent variable. An index was constructed for measuring to attitude of the change agents towards lab to land programme of the ICAR which is a technology transfer programme.

The index had 9 statements. Each of the statement was presented with a 4-point continuum, ranging from strongly agree to strongly disagree. The favourable statements were given scores for, strongly agree-4, agree-3, disagree-2 and strongly disagree-1. Scoring procedure was reversed for unfavourable statements. The content of the index for measuring the attitude was obtained by discussion with the scientists and experts implementing the lab to land programme and was considered to be valid. The index of 'Attitude towards Lab to Land Programme' is given in Appendix-I.

Independent Variables

The constraints perceived by the scientists and technologists in the transfer of agricultural technology under the lab to land programme were considered as independent variables. After collecting, editing and pretesting, 53 constraint items relating to management practices, infrastructure and work methods of the organisation, human relations, personal policies, union relations and social problems were finally selected.

To measure the degree of constraint, the respondents were asked to indicate a 4-point continuum the extent to which the constraints were perceived by them as hindering the technology transfer programme. The scoring procedure for the responses were, very much-4, much-3, not so much-2 and not at all-1.

Findings and Discussion

At the outset, correlation co-efficiencies between the dependent and 53 independent variables were calculated separately for the ICAR and non-ICAR system organisations. It was found that 18 variables were negatively and significantly correlated with the attitude of the change agents towards transfer of technology with at least one system of organisations.

To bring out an easily comprehensible simple structure of factors relating to organisational constraints, factor analysis of the 18 variables was done with the responses obtained from the total number of 40 respondents, comprising scientists and technologists.

Following (Harman Harry H. 1976) matrix of 18 independent variables was computed. Next, eigenvector matrix, eigenvalues and total variance percentage were calculated. In the present study, the eigenvalues which were greater than 1 were retained.

In respect of the independent variables, the following eigenvalues were greater than 1.

4.8037, 2.6583, 1.8602, 1.6631, 1.1976, 1.0980.

The total variance per cent explained by the eigenvalues were as follows.

26:69, 14.77, 10.33, 9.24, 6.65, 6.10.

The cumulative percentage of the eigenvalues were

26.69, 41.46, 51.79, 61.03, 67.68, 73.78.

Table 1 : Rotated Factor (VARIMAX) Matrix of Constraints Perceived by Change Agents

	Variables (constraints)	FACTOR						Communality
		1	2	3	4	5	6	
X_2	Tension in the organisation	.3910	–.1070	–.1307	.0090	–.1448	–.7788	.8089
X_6	Difficulty in procurement of inputs	.3837	.1614	.5502	–.2053	.1267	–.4040	.6974
X_9	Lack of free & informal communication amongst staff	.0072	–.0694	.7751	–.0231	.0012	.0253	.6068
X_{11}	Weak administrative structure of the organisation	.7858	.1236	.1929	.1825	.0860	–.2742	.7859
X_{16}	Lack of stability of the organisation	–.0999	–.2186	–.2556	.8207	–.0992	–.1962	.8449
X_{18}	Lack of influence of Govt. Department by the organisation	.5921	.3319	.4414	.0147	.2924	.1354	.7596
X_{25}	Complicated office procedure	.6893	–.3649	.0655	–.0363	.2422	–.1514	.6955
X_{26}	Lack of prompt decision by the Head of Organisation	.8132	–.2444	–.0099	.0725	.1443	–.0470	.7493
X_{28}	Lack of incentive in Technology Transfer Programme (TTP) work	.2259	–.2087	.5973	.5343	.0343	.0121	.7381
X_{29}	Lack of dignity in TTP work	–.2293	–.1225	.5687	.1415	.4159	–.0904	.5922
X_{30}	Interference with normal work	.0153	–.8456	.1304	.0523	.2185	–.0385	.7842
X_{33}	Lack of career advancement in TTP work	.1971	.0232	.7263	–.0205	.0942	.1273	.5924
X_{36}	Lack of clear-cut responsibility in TTP	.3397	–.1383	.0794	.0694	.7896	.0730	.7745
X_{37}	Lack of clear-cut instructions in TTP	.1637	–.0237	.2949	.0134	.7575	–.2102	.7325
X_{45}	Insufficient training of change agent on TTP	–.0424	–.1847	–.0013	.2061	.4162	–.6981	.7390
X_{46}	Lack of training aids in the organisation	.2696	.2597	.2234	.7617	.1881	.0454	.8077
X_{47}	Lack of technical know-how	.1788	–.5668	–.1709	–.1465	.5732	.0052	.7324
X_{53}	Affect social life of change agent	.1280	–.8928	.0463	.0622	–.0270	–.1398	.8398

For extraction of factors, the principal component method was adopted. The number of factors extracted from 18 independent variables was 6. The method of 'Varimax' rotation was used.

For interpretation of the factors, variables with high factor loading and high communality were taken into consideration. The six factors represented about 74 per cent of the total data variance. The rotated factor (Varimax) matrix of the independent variables is presented in Table-1.

Factor 1: Organisational System

Factor 1 accounted for 27 per cent of the total data variability. Three variables with high communality and high factor loadings were chosen to interpret this factor. The factor loadings were all positive. The variables and their factor loadings were, lack of prompt decision by the head of the organisation. .8132, weak administrative structure of the organisation .7858 and complicated office procedure .6893. These three variables were found to be positively and significantly correlated with each other.

Organisations are social units or human groupings deliberately constructed and reconstructed to seed specific goals (Etzioni 1976). Organisations have both structure and function. Change agents can not properly function in a rigid bureaucratic system.

Head of the organisation, generally, makes policy decisions which must be made in time and communicated to the change agents at the operational level. Lack of prompt decision by the head of the organisation, may vitally effect the technology transfer programme.

An administrative structure may be weak due to inadequate number and lack of competent personnel, inadequate finance and facilities etc. The change agents can not function properly in the organisation where the administrative structure and function are weak.

Bureaucracy has a tendency to complicate office procedure. Complicated office procedure may result in loss of time and energy and may develop conflict situation in the organisation. This is likely to hinder achievement of the goal for which the organisation has been set up. Change agents loose much of their operational efficiency when placed in such a situation.

Factor 1 was considered to be a general factor and was named as Organisational System.

Factor 2 : Quality of Working Life

This factor accounted for 15 per cent of the total data variability. Two variables were chosen for interpretation of the factor. The factor loadings were all negative. The variables, affect social life of change agent and interference with normal work had high loadings (-) .8928 and (-) .8456 respectively on this factor. These two variables were also found to be positively and significantly correlated with each other at 1 per cent level.

Implementation of the lab to land programme was an additional work for the change agents which involved long hours of working, sometime affecting their normal work and social life.

The factor may be termed as Quality of Working Life.

Factor 3 : Compensation

Factor 3 accounted for 10 per cent of the total data variance. On the basis of high factor loading and high communality, two variables were chosen to interpret the factor. The factor loadings of the variables were positive. The variables and their loadings were, lack of free and informal communication amongst staff .7751 and lack of career advancement in technology transfer work .7263. These two variables were found to positively and significantly correlated with each other.

Free and informal communication amongst staff members may help in problem solving and reducing tension. This may develop a cordial atmosphere in which the change agents would be able to contribute more towards technology transfer programme.

Career advancement in one of the basic motivations for job holders. In lab to land programme, there is no clear-cut instruction to give weightage for transfer of technology work during assessment for promotion of change agents. As such, the change agents may have a feeling that the transfer of technology work does not provide for sufficient incentive for career advancement.

In a bureaucratic set-up, the official receives regular pecuniary compensation of a normally fixed salary and old age security provided by a pension...The official is set for a career within the hierarchical order of the public service. He moves from the lower, less important, and lower paid to higher positions (Weber 1970).

Factor 3 was, therefore, termed as Compensation.

Factor 4 : Organisational Stability

This factor accounted for 9 per cent of the total data variability. The two variables which were chosen to interpret the factor were, lack of the stability of the organisation and lack of training aids in the organisation. The loadings of the variables were positive and were .8207 and .7617 respectively. These two variables were also found to be positively and significantly correlated with each other at 5 per cent level.

Stability of an organisation is an essential precondition for its growth and development. The non-ICAR system organisations, being voluntary in nature, are likely to be less stable. This was perceived as constraint mainly by the change agents of non-ICAR system organisations.

The training aids are considered important for technology transfer work. The organisations perceived as less stable are likely to have fewer training aids and, even if the training aids are available, they may not be in proper condition or less likely to be used.

On the basis of these variables, Factor 4 was named organisational stability.

Factor 5 : Authority and Responsibility

Factor 5 accounted for 7 per cent of the total data variance. Two variables were chosen for interpretation of this factor. The factor loadings of the variables were positive. The variable with their factor loadings were, lack of clear-cut responsibility in technology transfer programme .7896 and lack of clear-cut instructions in technology transfer programme .7575. These two variables were found to be positively and significantly correlated with each other at 1 per cent level.

Authority is the tool by which a manager is able to exercise discretion to create an environment for individual performance. Authority furnishes the primary line of communication in an enterprise, since it deals with those communications which are composed of decisions (Koontz 1984).

The Authority of the head of the organisation for technology transfer work is passed on to the change agents in the form of instructions. Unless the instructions are clear, they are not likely to produce the desired impact. It is also the job of the head of the organisation to see that the functionaries under him have clear-cut-responsibility. If, either the instruction or the responsibility is ambiguous the response is likely to be indifferent.

Factor 5 was termed as Authority and Responsibility.

Factor 6 : Organisational Climate

Factor 6 accounted for 6 per cent of the total data variability. Two variables were chosen to interpret the factor. The factor loadings of both the variables were negative. The variables with their loadings were, tension in the organisation (-) .7788 and insufficient training of change agent on technology transfer programme (-) .6981. These two variables were found to be positively and significantly correlated with each other at 5 per cent level.

Tension in an organisation indicates stress situation which may reduce its effectiveness. Insufficient training of the workers may also adversely affect the effectiveness of the organisation. These related to organisational climate on which, according to Koontz (1984) depends effective motivation of its functionaries.

The factor may be termed as Organisational Climate.

The constraints factors derived by factor analysis of the independent variables, perceived by the scientists and technologists hindering transfer of agricultural technology to the farmers are presented in Table 2.

Recommendations

1. Weak organisational system and lack of suitable organisational climate were found to be important factors hindering transfer of technology work by the change agents.
   Suitable orientation programmes may be organised for the head and other key functionaries of the organisation to ensure prompt decision and adoption of simplified and effective office procedure.
2. Change agents perceived that technology transfer work affected their quality of working life and there was lack of adequate compensation for this work.
   For proper motivation, there may be built-in procedure for career advancement of change agents in technology transfer work.
3. Lack of adequate authority and responsibility were found to be affecting technology transfer work of the change agents.

Table 2: Change Agent Constraints in the Transfer of Agricultural Technology

Variables	*Factors*	*Change Agent Constraints*
Lack of prompt decision by the head of the organisation Weak administrative structure of the organisation Complicated office procedure	1. Organisational System	Weak organisational system
Affect social life of change agent Interference with normal work	2. Quality of Working Life	Low quality of working life
Lack of free and informal communication amongst staff Lack of career advancement in technology transfer work	3. Compensation	Lack of adequate compensation
Lack of stability of the organisation Lack of training aids in the organisation	4. Organisational Stability	Lack of organisational stability
Lack of clear-cut responsibility in technology transfer programme Lack of clear-cut instructions in technology transfer programme	5. Authority and Responsibility	Lack of adequate authority and responsibility
Tension in the organisation Insufficient training of the change agent on technology transfer programme	6. Organisational climate	Lack of suitable organisational climate

Clear-cut instructions and commensurate authority and responsibility may be given to the change agents for proper implementation of the technology transfer programme. Periodical monitoring may also be arranged to ensure that the instructions are properly followed.

4. Lack of organisational stability as a constraint hindering technology transfer work was mostly perceived by the change agents of the voluntary organisations.

In sanctioning technology transfer programme, the capacity, competency and stability of the voluntary organisations may be taken into consideration.

References

Bennis, G. Warren., Changing Organisations, Bombay, Tata McGraw - Hill Publishing Co. Ltd., 1966.

Etzioni Amitai, Modern Organisations, New Delhi, Prentice - Hall of India (P) Ltd., 1976.

Harman Harry H., Modern Factor Analysis, Chicago and London, The University of Chicago Press, 1976.

Koontz, H., O' Donnell, C. and Weihrich, H., Management Japan, Mc Graw-Hill International Book Company, 1984.

Weber Max, Essays in Sociology, Edited by H.H. Gerth and C. Wreight Mills, London, EC 4, Routledge and Kegan Paul Ltd., 1970.

Appendix -I

Attitude Index of Change Agents Towards Lab to Land Programme

You will probably agree with some of the following statements of opinion about LLP and disagree with others. Please indicate your own feeling about each of them.

Sl. No.	*Statements*	*Strongly agree*	*Agree*	*Disagree*	*Strongly disagree*
(i)	Working for LLP is an additional burden				
(ii)	LLP helps to increase income of the adopted farmers				
(iii)	LLP is a stunt				
(iv)	Working for LLP does not count towards promotion				
(v)	LLP gives job satisfaction				
(vi)	It has been necessary to participate in LLP just for the request of Head of Division/Department				
(vii)	Working for LLP has little recognition				
(viii)	LLP provides opportunity for better extension work				
(ix)	It has been necessary to participate in LLP just to satisfy the Head of the Organisation				

15

Farmers Involvement in Research and Extension: Experience of Zambia

I.C. Nkhungulu

Background

Zambia is a landlocked country in Southern Africa lying in the elevated plateau area 8 degrees South and 18 degrees South latitudes (Fig. 1). Relatively it is a large country (about 753,000 sq km with a small population (7,153,700 by 1987). Despite a rapid growth rate of 3.1%, its overall population density is low (Table 1).

Table 1: **Basic Demographic Statistics**

	1969	*1980*
Population (Millions)	4.1	5.7
Male:Female Ratio	96.0	95.8
Population in Urban Areas (%)	29.4	43.0
Population density (persons per sq. km)	5.5	7.5
Crude birth rate (per 1000)	47.7	51.0
Crude death rate (per 1000)	19.7	16.6
Life expectancy, males yrs.	41.8	50.4
Life expectancy, females yrs.	45.0	52.5

Source: World Bank, 1987 and Government of the Republic of Zambia Central Statistics Office (GRZ CSO), 1985.

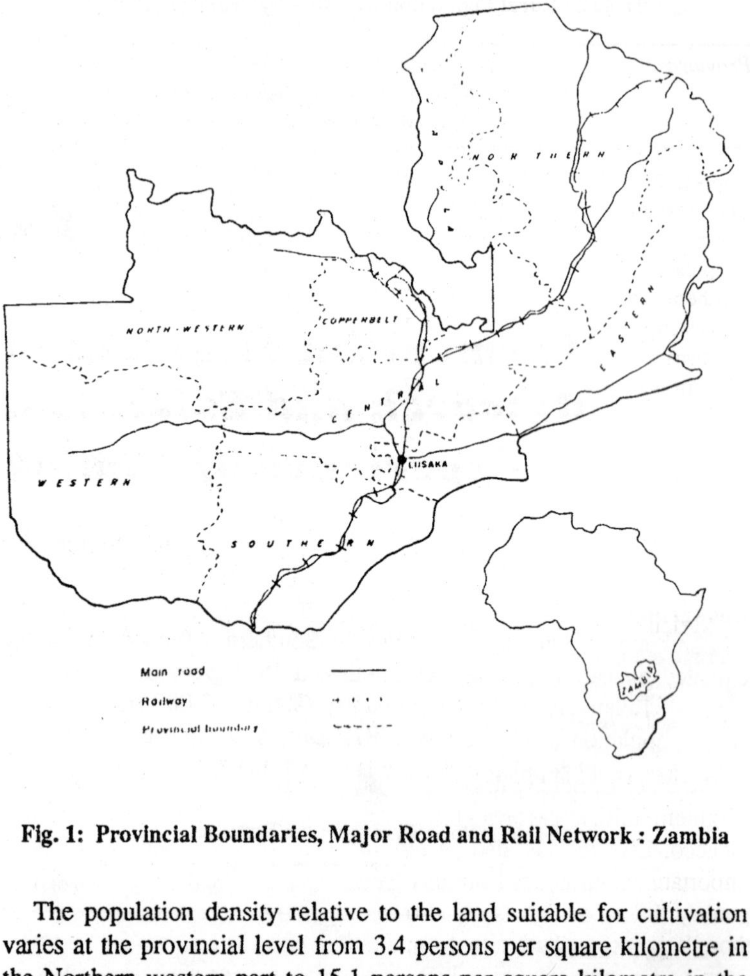

Fig. 1: Provincial Boundaries, Major Road and Rail Network : Zambia

The population density relative to the land suitable for cultivation varies at the provincial level from 3.4 persons per square kilometre in the Northern western part to 15.1 persons per square kilometre in the Eastern part (Table 2). Although the figures do not show any extreme pressures on the land, there are considerable variations within provinces. Parts of the Eastern and Southern provinces have experienced shortages of arable land for some years. Pressures on grazing resources also exists in some areas where successful small scale commercial farmers have accumulated wealth in the form of cattle.

Table 2 : **Rural Population Densities by Province, 1980**

Province	*Area suitable for cultivation (sq. km)*[1]	*1980 rural population*	*1980 persons per sq. km arable*
Copperbelt	23,735	1,00,034	4.21
Central and Lusaka	65,747	4,45,493	6.78
Eastern	39,498	5,97,038	15.12
Luapula	28,839	3,47,251	12.04
Northern	79,312	5,60,145	7.06
N/Western	78,368	2,63,632	3.36
Southern	45,608	5,00,386	10.97
Western	78,758	4,24,410	5.40
Total	4,39,865	32,39,389	7.36

1. This comprises state land, cropped land, and unused woodland.

Source: Agricultural Baseline Data for Planning—National Commission for Development Planning.

Agricultural development has been most significant since 1980 when average annual growth in agricultural GDP was 2.9% between 1980 and 1985. Nearly two-thirds of rural households are engaged in agricultural and the bulk of marketed agricultural produce is by Small Scale Commercial and Traditional Farmers.

Principal agricultural commodities in Zambia are maize, wheat, sorghum, millet, cassava, beans, groundnuts, soyabeans, cotton and tobacco. Livestock is also an important agricultural sector. The most important commodities from this sector are beef and dairy products and poultry. Sheep and goats are also reared mainly by small-scale farmers though their contribution to the market has so far been insignificant.

Farming Community

The farming community in Zambia is stratified into four categories based on farm size and degree of market orientation (Table 3).

Large-scale commercial farmers in Zambia are mechanized and use high technology inputs. They rely on permanent hired labour for farm operations. Area under cultivation varies between 40 and 5000 hectares. They cultivate maize, soyabeans, irrigated wheat, seed crops and horticultural crops and rear beef and dairy cattle and poultry.

Table 3 : Number of Farm Households by Farm Type

Farm Type	*Average Cultivated Area (Hectares)*	*Number of Farm Households*
Large Scale Commercial	> 40	730
Medium Scale Commercial	10 - 40	21,350
Small Scale Commercial	1 - 10	122,400
Subsistence	1 - 2	462,600

Source: Government of the Republic of Zambia, Ministry of Agriculture and Water Development (GRZ MAWD), 1981.

The Medium Commercial farmers market principally all of their produce. They use commercial technologies and rely on ox cultivation for draught power, together with family labour.

Small-Scale farmers rely mainly on hand cultivation supplemented, in some areas, by ox cultivation and some tractor hiring. This group produces about 60% of the marketed maize in the country.

Subsistence farmers, relying almost entirely on hand cultivation, predominate in Zambia (Table 2). They produce occasionally sell some produce.

Apart from farm size, another source of variability among farmers is gender. The 1980 population census indicated that in three Provinces, more than one-third of the rural households are headed by women. This has important implications for agricultural development in general and farmer involvement in research and extension in particular. A number of farmer organizations, which primarily represent the interests of large-scale farmers, represent members' concerns to the Ministry of Agriculture and Cooperatives (MACO) (formerly Ministry of Agriculture and Water Development [MAWD]).

Organizations in Technology Transfer

The two main institutions involved in technology transfer are Research and Extension Branches of the Department of Agriculture. Most of the agricultural Research is conducted by the Research Branch. Some research is also conducted by the Faculty of Agriculture in the University of Zambia, often in conjunction with the Research Branch.

Extension activities are the responsibility of the Extension Branch though some institutions also carry out some advisory activities. These include farmer organizations, churches, credit institutions, agricultural parastatal companies and marketing institutions.

While the Research and Extension Branches now look after the interests of all categories of farmers, this was not so prior to 1964. Previously, Zambia (when it was under colonial rule) had a dual agricultural system which looked at European (or Settler) and indigenous farming. The then European farming was directed at commercial production to satisfy beef and maize requirements in urban centres and in the copper mining areas. Agricultural research and extension for this sector was conducted by professional expatriates who worked mainly on commercial crops. Livestock did not attract much attention from the Department points of view though animal draught power was used for most of the farmers' operations until the late forties. Interaction between settler farmers and the Departmental staff was high resulting in strong linkages being developed between the two.

African, or indigenous, farming which was by far the largest group, mainly was involved in subsistence agriculture. Research and Extension efforts were directed at encouraging more permanent cultivation as opposed to the shifting cultivation being practiced then and development of commercial agriculture. The extension service was more authoritarian than persuasive.

After independence, in 1964, the two systems were merged into one under the Department of Agriculture. Since most of the Researchers were working on commercial crops, research continued on disciplinary lines at research stations under high management conditions. Little attention was paid to the small scale farmers' social and economic environment. After concern was expressed the research was not addressing Small Scale farmers problems, that Research Branch was reorganized with sixteen commodity and specialist Research Teams being established on an interdisciplinary basis to conduct on-station component research. In addition, Adaptive Research Planning Teams were established to operate in each of Zambia's nine Provinces to conduct farming systems research, including on farm trials under farmers' conditions.

Developments in Research Branch were also matched by similar moves in the Extension Branch. Extension efforts focussed increasingly on Small Scale farmers. Programmes aimed at addressing this category were introduced to try and hasten development. Apart from the conventional approach inherited from the earlier system, three other approaches were introduced:

— Integrated Rural Development Programmes which look at development of the rural areas to improve standards of living among the rural dwellers. Agricultural Extension Service forms just part of such an approach.

— Commodity based Extension Service has been introduced since the late 70's to encourage the production of some cash crops such as Cashew, Cotton, Coffee, Soyabeans and Tobacco. Under this approach, farmers are offered a complete package from inputs to marketing and advisory services.

— The latest approach is the modified Training and Visit System which was introduced in 1978. This has been partially introduced in three Provinces on a pilot scale. To complement the T & V system, a 'Lima' programme was introduced to facilitate the adoption of technological packages for Small Scale farmers. This programme encourages the use of agricultural inputs for the major cash crops on a standard unit, Lima (0.25 ha), of land using standardized measures of seed, fertilizer and insecticide.

Farmers Involvement in Research and Extension

The involvement of farmers in research and extension dates back to the colonial period. Settler farmers were incorporated in a number of research programmes using their farms and experience to enhance cash crop research. Mainly, maize and tobacco were, at the time, considered major cash crops, with other crops such as wheat and coffee being secondary in importance. Apart from the organized research conducted by the Department, farmers carried out what may be termed as crude research, where they imported several varieties of crops and livestock breeds and observed their performance under local conditions.

Indigenous farmers' involvement then was mainly as end users of technologies developed by the Department of Agriculture. As such, no research work was done in conjunction with the small scale farmers. However, due to the Government policy that the indigenous farmer undertake commercialized farming, some innovative farmers, after implementing certain technologies, modified them through trial and error to fit their circumstances. Once these modifications were in place, other farmers imitated the innovative farmers. So, indirectly, the innovative farmers acted as technology modifiers and extension workers.

After independence, in 1964, farmer involvement continued along the same lines. However, with the policy changes in the '70s for Research and Extension to cater mainly for Small Scale farmers, involvement of Small Scale farmers in the Research and Extension activities increased markedly.

Commercial farmer involvement has been both through direct involvement and private efforts. The Department of Agriculture has engaged some farmers in developing research strategies to address some of the problems experienced by farmers. Trials on wheat, soyabeans and maize have been set up on farmers[1] fields with the farmer providing management component and the Research Branch the technological component of the trials. Results from such trials are disseminated to other farmers through field days, farmer meetings and workshops. Privately, some farmers have undertaken research on such crops as sugarcane, tea, small stock and herbicide usage. Results from such trials usually are utilized by those farmers.

Small Scale farmer involvement has become paramount since the institution of ARPT and the T & V system. After ARPT have analyzed particular farming systems and pinpoint areas of technical intervention, trials are set up on farmers fields to address cultural and management aspects. Farmers are also involved in the baseline data surveys carried out to identify problems and bottlenecks experienced by Small Scale farmers. Farmers have also been involved in the feedback mechanism. Their contributions in meetings and workshops have assisted ARPTs to redefine, if necessary, their objectives.

In Extension, farmer involvement had been to some extent restricted to field days, demonstrations and training. Crop and livestock demonstrations were set up at interested farmers fields. This was so mainly during the time when commodity and conventional systems were in place. However, since the introduction of the Training and Visit System, farmer involvement has increased markedly. Contacts between Extension workers and farmers were initially individualistic with contact farmers (who were mainly male) acting as the link between other farmers and the Department. The participation however, was not very positive in spreading the technological messages probably because of a misunderstanding on whether they would be remunerated. It was also noted that mainly male farmers benefited most from the system though most of the work was done by women. Female headed households generally did not benefit from the technologies generated by the system. This approach to involve individual farmers in extension activities was replaced by a group approach. While previously most of the contact farmers were selected by the extension workers, the contact farmer group was selected by the farmers and this has made farmer involvement in extension activities more accepted. Women involvement also increased.

The Contact Farmer Groups in many cases comprise innovative, popular and representative farmers who are able to extend technological messages to fellow farmers. Innovative farmers provide the lead within a local community after experimenting informally with the new crops or other technologies made available by the Extension Worker. These come up with adaptive solutions suiting their locality. This modification then becomes adopted by other farmers with similar circumstances.

Farmer involvement in decision-making on demonstrations and field days has been quite varied. In settlements schemes, where farmers have been settled from densely populated areas, farmers make their own decisions on the running of such schemes.

Involving farmers in the programme helps to keep the Department on the right track. Farmers involvement in the Department has been useful in a number of ways.

Positive assessments have helped in developing technologies that farmers are able to use to their benefit. For instance, farmers commented favourably about several different maize varieties, both hybrids and composites, some of which are early maturing than material previously available. As a result of these comments, supported by statistical and economic analysis, these varieties have gone into demonstrations and widespread use.

Farmers have also commented positively about the "steadfast" variety of finger millet. It had larger heads than the local variety and was judged to be acceptable for food. The same goes for a bean variety "Carioca" which has been accepted, with management and cultural modifications agreed by farmers, and produced significant results.

Farmers have been instrumental in the modification of some of the recommendations made by the Department of Agriculture. For instance, they had questioned the wisdom of planting beans in flat seedbeds because of waterlogging and assisted the Department to develop technologies to plant beans on ridges. Farmers also disliked the maize/ bean intercropping, because beans planted between maize rows slowed weeding, necessitating its discontinuation. In the Eastern part of the country, farmers comments about weeding led the Department to change trials of simulated farmer practice from one to three weedings.

It must be noted, however, that not all farmers comments were agreed to by the Department of Agriculture. For example, the Research Branch had tested rice transplanting compared with the farmers' practice of broadcast seeding. At first, farmers dropped the recommended practice complaining that transplanting was too laborious. However, the Department continued with trials and the farmers later found that the advantages from transplanting were significant and interest has gradually increased.

Some negative assessments also helped the Department to discontinue some triais. When a poorly fabricated rotary jab planter was tested on farms in several Provinces, it gave highly variable seeding rate and depth, leading to the discontinuation of the trials.

References

GRZ CSO (1985). *Country Profile: Zambia 1985*. Central Statistics Office. Lusaka.

GRZ MAWD. (1981). *Food Strategy Study: Annex Socio-economic Characteristic of Farming in Zambia.* Planning Division, Lusaka.

Mulele, M.R. and Chungu, R.K. (1984). *Country Experiences of Extension and Research Linkages.* Paper presented at the African Workshop on Extension and Research, Kenya.

Kean, S. and Singogo, L.P. (1988). *Organization and Management of Adaptive Research Planning Teams in Zambia.* ISNAR Publication. The Hague.

Silangwa, S.M. (1982). *Linkages between Extension Services and Research and Experimental Development Activities.* Paper presented at the Symposium on the Role of Extension Services in National Development. Lusaka.

Sutherland, A. (1987). *Agricultural Extension and Farming Systems Research in the Field.* A Case Study from Zambia. Department of Agriculture.

Part 4

Training and Technology Transfer

"A person will learn when a goal is realised, dissatisfaction with his or her own performance is felt, or the person wishes to achieve certain socio-psychological or economic rewards. In offering opportunities for individual or group growth, more and more training and development efforts are focused on the use of self-motivation principles as the achievements and the solution of problems best motivate people ....."

Gordon L. Lippitt, 1982.

16

Training of Extension Personnel for Effective Technology Transfer

Blannie E. Bowen and Harry A. Carey

Agricultural extension agents have long enjoyed an enviable position of influence within the agricultural community. These professionals fulfil numerous roles, but perhaps the most crucial is that of a change agent in helping farmers adopt new technology. Extension agents are commonly viewed by farmers with trust and respect for three primary reasons. First, extension agents are usually associated with reputable land grant universities. Second, they reflect current research findings. Third, extension agents provide unbiased yet high quality information and services. In addition, even though agriculture experienced numerous changes during the 1980s, extension agents still enjoy the traditional high level of respect within the agricultural community.

Other than trusted family members, no individuals enjoy more credibility with the farm community than county level agricultural extension agents. From the perspective of farmers, county extension agents rank second in terms of providing credible information and advice. When they make production and marketing decisions, farmers feel confident about the caliber of assistance they receive from extension agents. Farmers also rank the agents very high in terms of importance and how often they use the information and services that extension provides (Kramic, 1987). Agricultural extension agents and other individuals involved in agriculture are ranked in Figure 1 according to their credibility.

When the farmers perceive an agricultural resource person to be an important source for production and marketing information, they also

Fig. 1 : Credibility of Agricultural Resource Persons

(Importance Farmers Place on, Their Use of, and Confidence in When Making Production and Marketing Decisions)

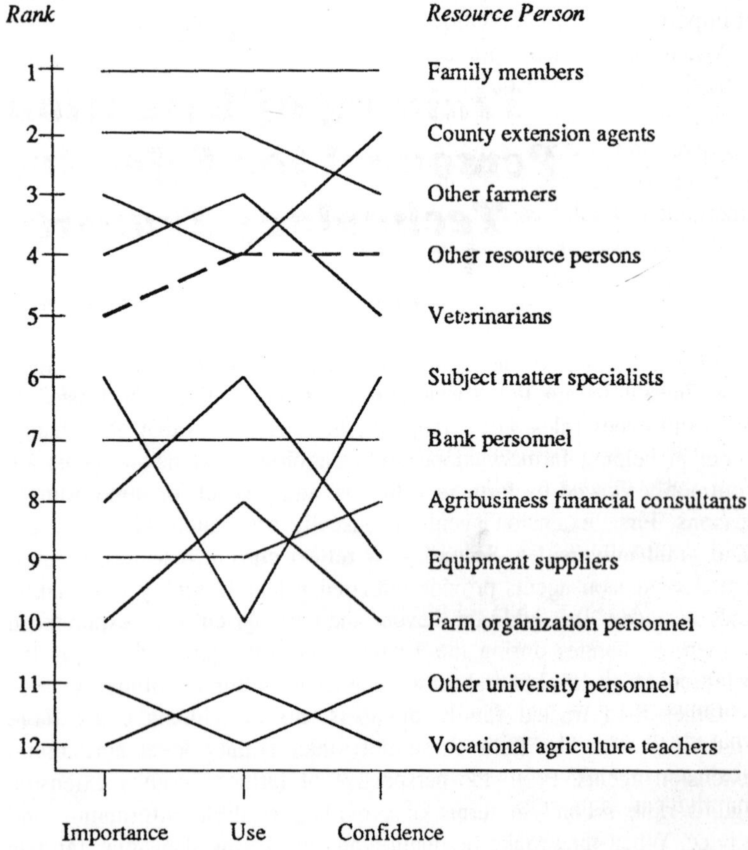

Source: Kramic (1987) Study of Ohio farmers.

tend to use that individual and express a great deal of confidence in the individual. A notable exception to this pattern is shown in Figure 1. Farmers rank extension subject matter specialists very high on both importance and confidence. However, the nature of extension as a delivery system precludes extensive direct contact and use of specialists by farmers. With this one exception, relationships among the concepts of importance, use, and confidence are realistic, not theoretical.

When farmers make production and marketing decisions, they rank extension agents high on all three concepts. Such rankings mean that extension agents are viable links in the diffusion and adoption process. However, to maintain credibility with the farm community, agents need high caliber communications and methodology skills to transfer emerging technologies from research and development laboratories to productive uses on farms and ranches. Extension administrators feel this is a global concern. In fact, Sigman and Swanson (1984) found that three of the top five problems facing national agricultural extension in developing countries involve communications skills and methodology:

1. "Extension personnel lack essential teaching and communication equipment" (p. 5).
2. "Extension personnel lack training in extension methods and communication skills" (p. 4).
3. "Extension personnel lack essential teaching aids, bulletins, demonstration materials, etc." (p. 5).

AGENT AND FARMER COMMUNICATION

A host of media and methods are used by an extension agents to help communicate new and emerging technologies to farmers. No matter what methods and media an extension agent uses to communicate messages to farmers, those messages usually flow through a formal communication process.

The Communication Process

An agent encodes verbal, nonverbal, written, and audio-visual messages that are intended for farmers. As shown in Figure 2, external distractions and internal differences occur in the form of static or interference to inhibit a farmer from correctly decoding and internalizing the messages. However, once a farmer decides on a response to the agent's message, this person encodes the response which

Fig. 2 : Agent and Farmer Communication Process

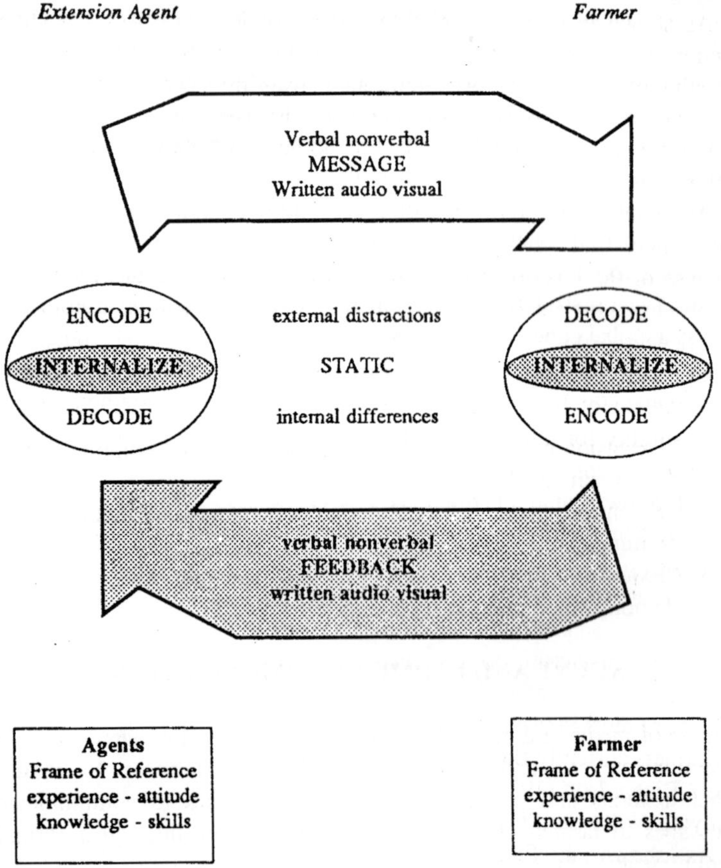

Source: Ridenour, H.E., 1977, p. 7.

can be in verbal, nonverbal, written, or audio-visual formats. As with messages transmitted by the agent, feedback from a farmer will also be influenced by static forces that perhaps impeded the agent's original message.

As shown in Figure 2, once the agent receives feedback from the farmer, this message must be decoded and internalized to grasp the full meaning of the message. An additional barrier influences both the level and type of communication that can occur between an extension agent and a farmer. This inhibiter occurs because of frame of reference differences.

An agent and farmer differ in their experiences, attitudes, knowledge, and skill. For example, experience differences occur because of the level of formal education an agent receives. The agent usually has a higher level of education, thus, broader and deeper formal experiences. While fulfilling professional responsibilities, these experiences should prove invaluable. Because the agent must serve in the capacity of a catalyst for change, this individual must assume responsibility for initiating and fostering two-way communication. An agent who wants to be effective in transferring technology to farmers cannot allow frame of reference and related differences to inhibit effective communication between the two parties.

Farmers and Extension Methods and Media

Extension agents often use a variety of methods and media without carefully considering to communication styles and preferences of farmers. Which methods and media used by agricultural extension agents are most important to farmers? Which of the methods and media do farmers currently find most useful and which will they use five years later? Richardson (1989) provided answers to these three questions by studying North Carolina farmers who used extension.

As shown in Figure 3, farmers ranked newsletters first in importance, current use, and future use, thus, making newsletters their preferred medium. Even though extension meetings ranked 3rd in importance and 2nd in current use, in the future farmers plan to rely on meetings to the same extent as extension newsletters. Farm visits, no-farm tests, and telephone calls are other methods and media that farmers will use extensively in the future. Ranking lowest in terms of future use are radio, fact sheets, videotape, fairs, and exhibits.

With a few exceptions, the importance that farmers placed on the methods and media tended to match their level of current and future

Fig. 3 : Use of Extension Communications Methods and Media by Farmers

(Importance Farmers Place on and Their Current and Future Use of 21 Extension Communications Methods and Media)

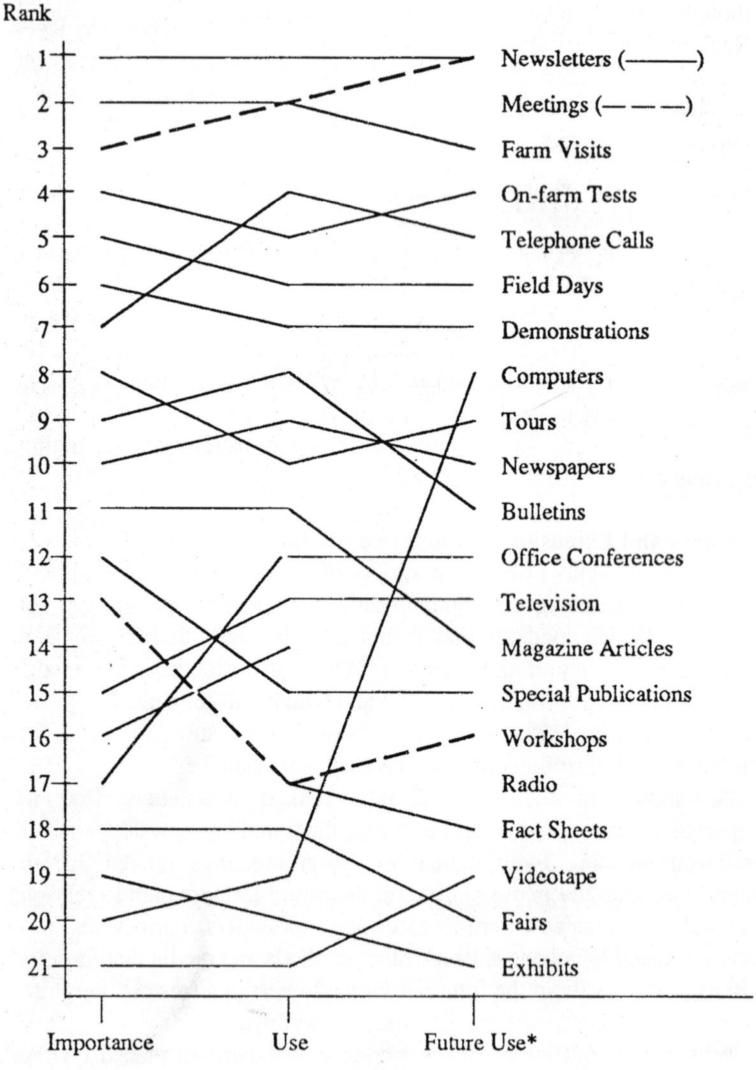

*Used the most 5 years from now.
Source: Richardson, 1989.

use. Computers ranked 20th in importance and 19th in current use, however, the farmers ranked this medium 8th in future use. Three other less radical departures from the importance - current use - future use match are also shown in Figure 3. Telephone calls and office conferences were used extensively and both will be in the future even though farmers attach a low importance ranking to these methods. Ranking high in importance, but low on current and future use were special publications and fact sheets. Bulletins, radio, and magazine articles ranked much lower in future use than either their importance or current use rankings.

AGENTS AS EFFECTIVE TEACHERS

An extension agent must have a solid core of foundation experiences that are focused around eight topics if that individual is to function effectively as a change agent. Included as foundation experiences are communications, professional, technical, and interpersonal skills. The Phipps and Osborne (1988) triangle presented in Figure 4 graphically presents the desired foundation experiences.

Fig. 4 : **Foundations of Becoming a Good Teacher of Agriculture**

Source: Phipps & Osborne, 1988, p. 138).

Effective Agent Behaviors

Once an agent acquires the foundations and teaching methods skills, what distinguishes effective extension agents from those who are not effective in having clientele adopt new innovations? Perhaps the most definitive answer for this often perplexing question comes through the research of Rosenshine and Furst (1971). These researchers synthesized the mammoth volumes of educational research about teacher effectiveness.

Numerous factors have been found to provide partial answers to why students learn. However, five factors that Rosenshine and Furst identified tend to be the best and most consistent explanations of why students learn. Not surprising, most of the factors relate to actions of the teacher. In effect, it is a teacher's behavior, i.e. what the teacher says *and* does while directing learning experiences, that best explain how much learning will occur. The Rosenshine and Furst behaviors include:

- *Business-like:* Common indicators of this behavior include starting and ending meetings at the appointed time, demonstrating full command of the learning environment, being the leader of the instructional process, keeping clientele sharply focused on the objectives of the session or meeting.
- *Clarity:* Providing concise, precise, and correct instructions without unnecessary confusion. There are no doubts about the goals and intended results. Time is not wasted on details that should be handled during a planning or preparation stage. Unrelated materials are not presented, thus reducing confusion.
- *Enthusiasm:* Approaching the task with inspiring zest, zeal, and charm; being able to transfer your excitement and interest in the topic to the appropriate clientele groups. Appropriate strategies are used to avoid boredom. Enthusiasm tends to be catching, even in an instructional situation.
- *Humor:* Being able to accept and give small-talk, appropriate stories; having a sense of humility and humbleness even when a joke or story might cast you in a somewhat unfavorable light; humor is used appropriately and is not permitted to delay or divert the instructional process.
- *Variability:* Using a variety of methods, media, and resources. Appropriate changes of pace are used to break monotony and boredom. Breaks, refreshments, and similar strategies are used during the instruction.

Listed in Figure 5 are the Rosenshine and Furst behaviors and interrelationships among the characteristics. An effective extension agent should exhibit these characteristics when delivering instruction. The extent to which clientele adopt innovations or technology will be associated with the effectiveness of the instruction an agent delivers.

Fig. 5 : **Effective Extension Agent Behaviors**

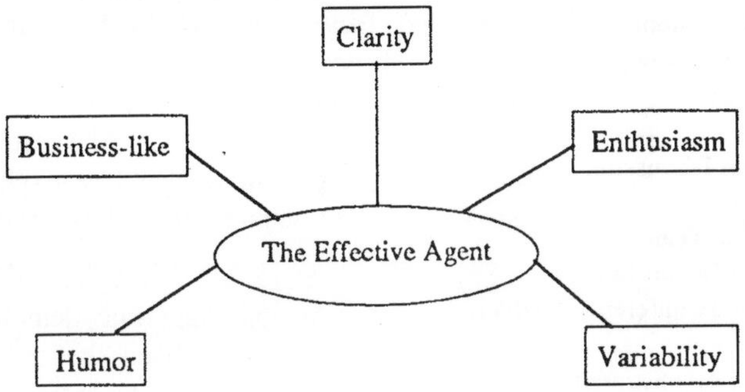

Source: Based on teacher effectiveness research by Rosenshine, B. & Furst, N. (1971).

Extension Teaching Methods

In addition to acquiring the Phipps and Osborne (1988) foundation experiences, an extension agent must also be knowledgeable of and proficient in using universally accepted teaching methods. Extension agents traditionally use a variety of methods and techniques while teaching farmers and other potential adopters about an emerging or different technology. As shown in the Table 1, Kang and Song (1984) have categorized the most common extension teaching methods as either a group or an individual technique.

The techniques are often provided to extension agents during a variety of undergraduate courses. Instructional materials, communications media and methods, and methods of teaching are common courses where the methods are taught. However, for agents who are hired without the desired methodology skills, a variety of inservice approaches will be necessary. Any approaches that are selected must be implemented with the understanding the developing educationally sound skills requires considerable investments of time, money, and related resources.

Table 1 : **Selected Individual and Group Extension Teaching Methods**

Group Methods	*Individual Methods*
Brain Storming	Farm and Home Visits
Buzz Session	Field Flag
Clinic	Informal Contacts
Colloquy	Model Farmer
Contests	Office Calls and Inquiries
Critical Incident	
Discussion Group	
Field Day	
Field Trip	
Informal Discussion	
Lecture	
Listening Team	
Method Demonstration	
Modified Conference Method	
Panel	
Result Demonstration	
Role-Playing	
Seminar	
Simulation Games	
Symposium	
Workshop	

(*Source*: Kang & Song in Swanson, 1984, 130-142.

Using Extension Methods and Media

Which of the Kang and Song (1984) methods do extension agents use most often? Methods and media that Ohio Cooperative Extension Service (OCES) agricultural agents used most often during 1987-88 while delivering instruction to farmers are presented in Figure 6. Office calls, telephone calls, bulletins, and newsletters were the methods and media used most often by OCES agents. The approaches used least were the mass media (radio, television, and magazines) and teleconferencing.

Fig. 6 : Communications Methods and Media Used by Agricultural Extension Agents

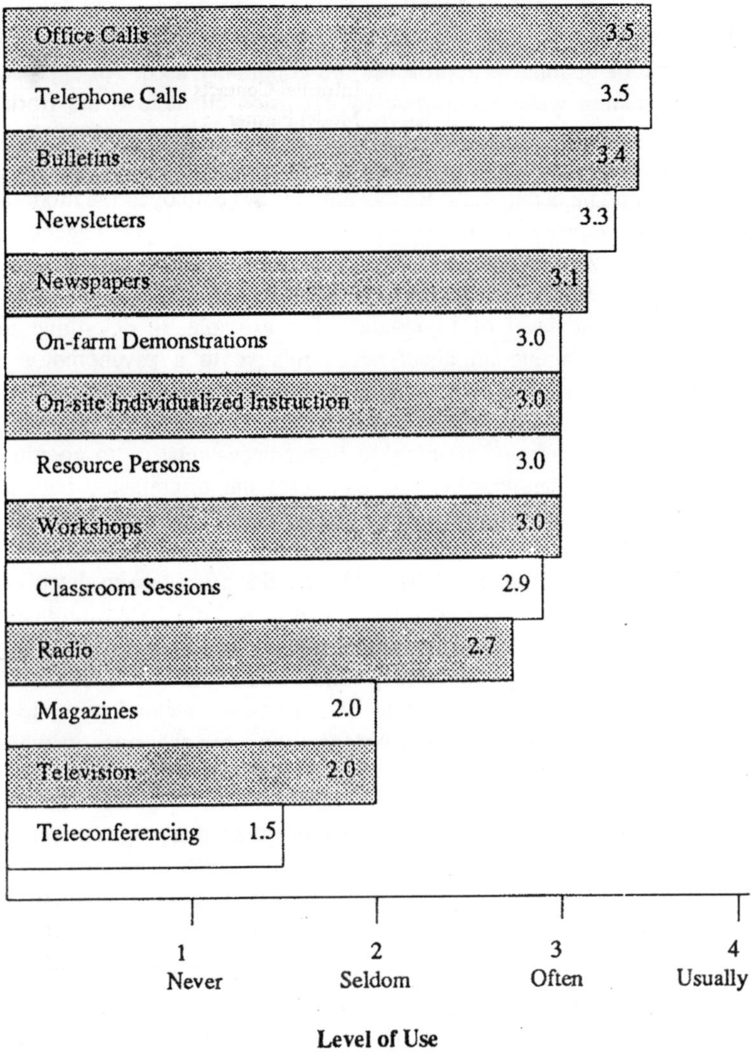

Source: Bouare (1988) Study of Ohio Cooperative Extension Service Agents.

ASSESSING COMMUNICATIONS AND METHODOLOGY NEEDS

Numerous approaches are available to determine which communications skills and teaching methods an agent needs to develop. Personal interviews, observations, self-assessments, questionnaires, and a variety of qualitative approaches are commonly used. An approach that is gaining wider acceptance by extension clientele is the Borich (1980) Discrepancy Model of Needs Assessment. Bowen (1988) used the model to determine the financial management training needs of extension home economists. Rennekamp (1987) employed the model to determine the professional knowledge and skills of 4-H youth agents.

The underlying logic or rationale for the model is that you can determine the difference between "*what is*" and "*what ought to be*" as a function of the level of importance. For example, to determine the photography training an agent needs relative to a psychomotor or performance skill, one computes the discrepancy between the ability to use a camera as a function of how important the skill is for all agents. Likewise, to determine training an agent needs relative to cognitive knowledge of photography, one computes the discrepancy between knowledge of photography and the relative importance of that knowledge.

Scherer and Masiclat (1988) used a similar procedure to determine the types of communications training that New York Extension agents needed. These researchers identified training needs in terms of new technologies, conceptual areas, and traditional communications methods. As shown in Figure 7, computer technology, video, communications strategies, and photography were the areas in which the agents had a moderate to high need for training.

INSERVICE TRAINING TECHNIQUES AND APPROACHES

Once the communications and instructional methodology needs of extension personnel have been determined, appropriate delivery methods must be used. Should the training be in combination with another type of training? Who is best qualified to deliver the training? Of what duration and at what level should the training be? Additional concerns relate to the literacy level of the extension personnel to be trained as well as the level for clientele who are to be served by extension professionals. An additional concern focuses on social and

Fig. 7: Per Cent of Agents Indicating Interest in Communication Training

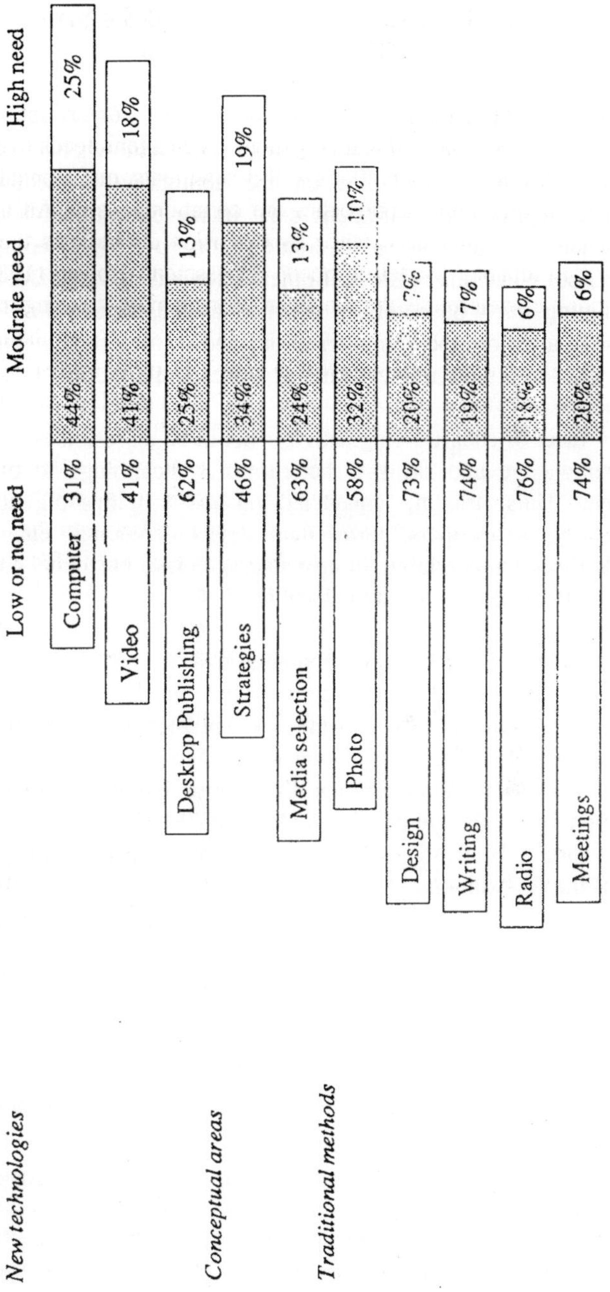

Source: Scherer and Masiclat (1988, p. 17) study of New York Extension Agents.

cultural expectations of both extension professionals and the targeted clientele groups.

Matching Literacy Levels

Literacy levels are often the crucial factor in whether or not a training session is conducted on the appropriate level and with sufficient depth. Wide variations exist in the educational levels and experiences of extension agents. Some agents have barely completed a high school education while still others have a doctorate. The level of the training must be carefully matched to the literacy level of the agents. In addition, the communications and instructional methods and media chosen for the training must also match the literacy level of the participants.

One facet of each training activity should be a session on using the new knowledge and skills to help transfer technology to extension clientele. Thus, training conducted on radio, television, computer technology, newspapers, newsletters, teleconferencing, etc., should include discussions relative to uses that extension personnel can make of the methods and media at the local level.

Matching Social and Cultural Expectations

Training conducted without appropriate consideration of the social and cultural contexts in which the new knowledge and skills will be applied tends to have limited long term effectiveness. The following social expectations relative to the diffusion and adoption process were identified by Lionberger and Gwin (1982, p. 75). Training for extension agents should be conducted with appropriate consideration given to each of the expectations:

* village	* work group
* friends	* family
* community	* cooperative
* farm organization	* religion

Lionberger and Gwin also indicate that the above social groups share common characteristics (p. 75):

1. The groups are comprised of individuals who associate and belong to the group.
2. Members of the group know and trust each other.
3. The group establishes thoughts, beliefs, and actions as group norms.

4. The group controls the actions of group members via rewards and punishment.
5. Members aid fellow members when need arises.

A Model for Transferring Communications Skills

Once the communications and instructional methodology skills an agent needs have been identified, appropriate delivery procedures must be planned and implemented. Listed below are philosophical, delivery, and evaluation considerations relative to providing communications and methodology training for extension agents:

Philosophical Considerations

What is the responsibility of the individual agent to provide self-improvement activities?

What role should the extension organization have into providing the needed training?

Are the needed skills job specific or of a self-improvement nature?

Are the extension organization's hiring procedures sufficient to screen for individuals who can bring the needed skills with them to the organization?

To what extent should professional improvement be linked with the organization's system of promotion and tenure?

Are incentives such as monetary rewards and advancement opportunities needed for agents to acquire the desired skills?

Delivery Considerations

What types of individual instruction are appropriate?

-Self study packets -Computer, audio, and videotape programs
-Tutorial packages -Learning activity packages

What types of group instruction might be appropriate?

-Workshops
-University courses (graduate or undergraduate)
-Other professional development activities

Who is most qualified and available to provide the training?

-Agricultural communications specialists -Extension training officers
-Subject matter specialists -Communications faculty
-Agricultural and extension education faculty -Industry representatives

Who must provide the equipment, publications, resources materials, supplies, and other items needed to conduct the training?
When is the appropriate time of the year to provide the training?
How should the training be sequenced?

Evaluation Considerations
How successful was the program in delivering the desired training?
How satisfied are the participants with the quality of the training?
Will the training be effective both in the interim and over a sustained period?
When similar training is provided, what must the organization do differently?
Was the training provided in a cost effective manner?
Does the extension organization have an on-going staff development plan for agents?
Will a self-evaluation program enable agents to become more involved in high quality professional improvement activities?

A Model Communications Workshop

A 5-phase communications skills workshop conducted for new Mississippi Cooperative Extension Service agents is shown in Figure 8. The workshop was conducted over a 15 month period and included many of the considerations noted above (Bowen, 1983). The workshop was planned and delivered by an agricultural and extension education faculty member at Mississippi State University.

COMMUNICATIONS TRAINING VIA GRADUATE STUDY

If an extension professional decides to pursue communications training through graduate study, two philosophical questions must be addressed immediately: (1) Am I seeking to become more informed about communications theory? Is my primary intent to become more skilled as a communicator? Answers to these questions lead to vastly different paths.

Graduate study is designed primarily to develop a thorough philosophical background and cognitive abilities rather than psychomotor skills. This dichotomy exists because in most graduate level courses students are expected to study, consume, and/or produce communications oriented research. Fortunately for students who wish to acquire communications skills through a master's degree, there are a

Fig. 8: Conceptual Model for 5-phase Communications Skills Workshop

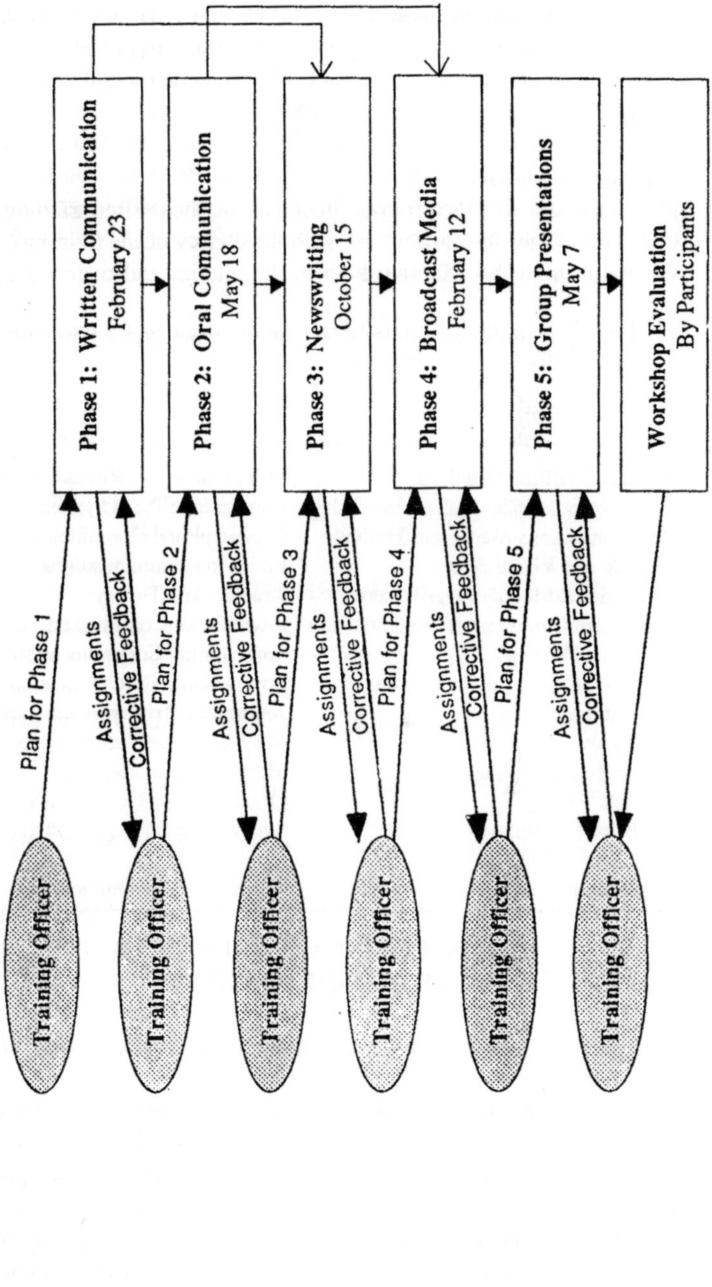

number of options. Most universities have courses that can be taken for master's degree, thére are a number of options. Most universities have courses that can be taken for master's degree or advanced undergraduate credit. Such courses usually provide the best alternative for securing, upgrading, or expanding existing communications skills. Advanced graduate and doctoral level courses tend to focus on communications processes and theories. Listed in Table 2 are sample graduate level communications and journalism courses offered by several universities that focus upon skills or theory.

Table 2 : **Sample Graduate Level Communications Courses Offered by U.S. Universities**

Skills	*Theory*
Advanced Editing Practices	Communications Research Methods
Communications/Journalism Internship	Comparative Press Systems
Communications Media and Methods	Cross-Cultural Communication
Graphic and Visual Arts	Ethics in Communications
Instructional Media in Agriculture	Graphic Arts Theory
Journalism/Communications Practicum	Intercultural Communication
Magazine Writing	International Communication
News Media and Public Opinion	Interpersonal Communication Theory
Photography	Journalism History in America
Press Law	Mass Communications Theory
Public Relations Practices	Mass Media and Society
Science Writing	Organizational Communications
Skills for Environmental Communications	Science, Technology, and the Media
Typography	Theories of Communication

TRANSFERRING COMMUNICATIONS RESEARCH TO AGENTS

A variety of individuals conduct research about processes and dynamics involved in transferring technology from research laboratories to consumers and users of that technology. As opposed to subject matter areas such as agronomy or horticulture where laboratory experiments can be conducted with maximum control and precision, processes involved in the adoption of new technology cannot be easily studied with the same degree of precision. Hence, a host of social science disciplines come into play when one seeks answers to questions related to the adoption of new technology.

Who Should Do the Needed Research

Lionberger and Gwin (1982) illustrate the intricacies involved in what Rogers and Shoemaker (1971) categorized as the diffusion and adoption process. In summarizing research and new knowledge generated about the process, Lionberger and Gwin developed nine categories that focus on educational strategies, individuals, and groups:

"1. The individual adoption process.
2. The function of information sources in the adoption process.
3. Diffusion curves; the nature of the community adoption processes.
4. The role of different types of people in the individual adoption process.
5. The importance of groups in the diffusion of information and the adoption of practices.
6. The influence of attributes of an innovation on its adoption.
7. The way farmers view information sources.
8. Influences on early and late adoption
9. Conditions that either retard or speed-up adoption" (1982, p. 60).

The above categories illustrate that a number of social scientists from a variety of disciplines actively seek answers about the types of forces and factors that influence the adoption of a new technology. Included are rural sociologists, communications and journalism faculty, agricultural educators, economists, psychologists, and a host of other individuals from a variety of social science disciplines.

Communications Research Model

Patterns followed in conducting, translating, and disseminating communications research are shown in Figure 9. Problems of a communications nature are sometimes identified by extension agents. These professionals then communicate such problems to agricultural communicators who often relay the problems to researchers who find solutions. Once the solution is discovered, it is transmitted to agricultural communicators via channels such as research and professional journals and through professional meetings. Agricultural communicators then employ a variety of methods and media to transmit the new knowledge to extension agents. An atypical communications research pattern is also depicted in Figure 9. On occasion, extension agents identify communications oriented problems and transmit them directly to researchers. Solutions to the problems are returned directly to the agents. This atypical pattern is most prevalent when extension agents are enrolled in graduate courses where they can conduct thesis, dissertation, or special problems oriented research.

Fig. 9 : Theoretical Communications Research Dissemination Model

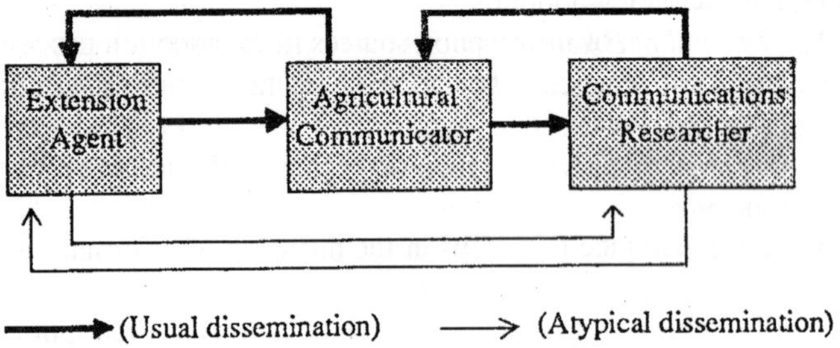

Who Transfers Communications Research

When new knowledge is generated about processes and procedures that can be used to effectively communicate a new technology to farmers, consumers, and other target groups, questions usually arise about who should transfer the new knowledge to agricultural extension agents, i.e. change agents. One often erroneously assumes that creators of a new technology will fulfil this role. However, history has proven that this rarely occurs. In most extension programs, this responsibility typically rests with staff development personnel, training officers, or the agricultural information department. These professionals must locate the diffusion and communications oriented research, translate it into practical knowledge, and then transmit that knowledge to extension personnel on all levels, i.e. state, district, and county levels.

How Must Transfers Occur

A variety of methods and forums are used to transfer new communications knowledge to extension agents. New agent orientation is an opportune time to disseminate communications research to neophyte extension professionals. This forum is especially suited to disseminating new practices and procedures so all new employees can commence their employment with a common base of experience. For veteran extension professionals, annual conferences, season updates, and technical training sessions offer appropriate opportunities to disseminate new communications practices and procedures.

For extension professionals who subscribe to the concept that individual agents share responsibility for staying current with new communications knowledge, a host of college courses and industry sponsored workshops, training sessions, and meetings are available. In addition, a wealth of printed materials exists. However, perhaps the best source for written materials about new communications knowledge is found in professional journals.

The *Journal of Extension* publishes research based articles where the focus is on practical applications rather than theoretical or methodological issues. This journal also has a special section called "Research in Brief" which focuses on research that may not reach extension professionals through other media. *ACE Quarterly*, the professional journal of Agricultural Communicators in Education, an international society of agricultural communicators, publishes original research, opinion pieces, how-to-articles, reviews, surveys or trends pieces, and analyses that are of interest to extension communications specialists. Both of these sources offer excellent materials that will be of interest to extension professionals of all levels within the organisation. For extension professionals working on the state or national levels, *The Journal of Agricultural Education* (formerly the *Journal of the American Association of Teacher Educators in Agriculture*) is an excellent source for research and philosophical articles about extension methods and media.

SUMMARY

Comprehensive and futuristic programs must be in place to enable extension personnel to effectively transfer technology to farmers and other clientele groups. The following recommendations are offered to help in bringing about this goal.

Preferences of Farmers

Extension methods and communications media that extension agents use in having farmers adopt technology must be appropriate to the needs of the target audiences.

Needs of Agents

A variety of approaches must be used to accurately and continuously assess the extension methods and communications needs of agents. Both qualitative and quantitative approaches must be used. Further, agents must value and have a vested interest in completing the intended training.

Providing Agent Training

The responsibility of providing extension methodology and communications training must be a shared undertaking. Even though extension communications specialists will have a major role in providing the training, staff development and subject matter specialists must also play significant roles in the overall training program. From a motivational perspective, agents must concur that they have the most significant responsibility for securing and using extensions methods and communications media.

Transferring Communications and Methodology Research

Extension communications and staff development specialists should have primary responsibility for transferring communications and methodology research to local extension agents. These professionals are usually closer to the original source of the research and are in ideal positions to translate the new knowledge into formats that are appropriate for agents. This is especially true for research involving new communications and instructional technologies.

References

Borich, G.W. (1980). A needs assessment model for conducting follow-up studies. *Journal of Teacher Education*, 21(3), 39-42.

Bouare, D. (1988). *Perceptions of adult educators regarding the content, delivery systems and types of farmers served by three young/adult education programs in Ohio*. Unpublished master of science thesis, The Ohio State University, Columbus.

Bowen, B.E. (1983, May 20). Enhancing communications skills of local agents. Paper presented during the National Extension Staff Development Conference, St. Louis, MO.

Bowen, C.F. (1988). *Financial management training needs of extension home economists*. Unpublished doctoral dissertation, The Ohio State University, Columbus.

Kang, J.T. & Song, H.K. (1984). Individual and group extension teaching methods. In *Agricultural Extension: A reference manual*, 2nd Ed., Swanson, B.E. (Editor). Rome: Food and Agriculture Organisation of the United Nations.

Kramic, J.L. (1987). *The level of impact of agricultural information sources on production and marketing decisions of Ohio farmers*. Unpublished master of science thesis, The Ohio State University, Columbus.

Lionberger, H.F. & Gwin, P.H. (1982). *Communication strategies: A guide for agricultural change agents*. Danville, IL: Interstate Printers & Publishers, Inc.

Phipps, L.J. & Osborne, E.W. (1988). *Handbook on agricultural education in public schools*. Danville, IL: Interstate Printers and Publishers.

Rennekamp, R.A. (1987). *The relationship between selected antecedent characteristics and the perceived educational needs of Extension agents with 4-H youth development responsibilities*. Unpublished doctoral dissertation. The Ohio State University, Columbus.

Richardson, J.G. (1989). Extension information delivery methods: Detecting trends among users. *ACE Quarterly*, 72 (1), 23-27.

Ridenour, H.E. (1977). *Selection and use of instructional resources*. The Heart of Instruction Series #4. Columbus, OH: Department of Education, Division of Vocational Education, and the Ohio Agricultural Education Curriculum Materials Service.

Rogers, E.M. & Shoemaker, F.F. (1971). *Communication of innovations: A crosscultural approach* (2nd. ed.). New York: Free Press.

Rosenshine, B. & Furst, N. (1971). Reseach on teacher performance criteria. In *Research in Teacher Education*, (Edited by B.O. Smith). Englewood Cliffs, NJ: Prentice-Hall.

Scherer, C.W. & Masiclat, S. (1988). Conceptual, traditional and new media training needs: A study of county extension agent perceptions. *ACE Quarterly*, *71* (4), 15-19.

Sigman, V.A & Swanson, B.E. (1984). *Problems facing national agricultural extension in developing countries*. Urbana, IL: The University of Illinois, International Programs for Agricultural Knowledge Systems (INTERPAKS).

17

Training of Extension Personnel and Farmers for Effective Technology Transfer

Md. Monirul Islam

The Concept of Training

Training is the process of improving the knowledge and skills and of changing the attitude of an individual for doing a specific job. In any organization people are one of the most vital inputs and they are considered as the greatest asset. Extension organizations deal with transfer of technology for the benefit of rural people. In this process the people in the organization must know how to make effective use of the available technology. They must be equipped with necessary knowledge, skills, attitudes, and abilities to use the technology with a view to bring about a continuous improvement in the quality of their work performance (Tylor, 1961). This can be achieved through appropriate training programmes. Training is, thus, concerned with the meeting between two inputs to organizational effectiveness: people and technology. Before the people can perform their tasks properly, they must master the special technology used by the organization (Laird, 1985). Thus, training can simply be defined as the act of increasing the knowledge and skills of an employee for doing a particular job (Flippo, 1976).

Education and Training

Although the words "training" and "Education" are sometimes used interchangeably, some distinction can be made between them. While training is concerned with "those activities which are designed to improve human performance on the job the employee is presently doing or is being hired to do (Nadler, 1970), education is concerned with "increasing general knowledge and understanding of our total environment (Flippo, 1976). Education is the development of human mind and it increases his power of observation, analysis, integration, understanding, decision-making, and adjustment to new situations. According to Nadler, education is those human resource development activities which are designed to improve the overall competence of the employee in a specified direction and beyond the job now held." In training the emphasis is on development of skills, but in education the emphasis is on broadening the outlook and horizon of knowledge. Educational activities are more general than training activities which are specific. However, these two terms can be placed at the two ends of a continuum and the personnel development activities be judged whether it is a training or education. Both training and education are involved in the development of personnel to the desired level of skill, knowledge, and attitude (Flippo, 1976).

Importance of Training for Technology Transfer

The importance of proper training to extension personnel and farmers can never be over-emphasized. In an extension organization various categories of personnel perform their functions with the prime objective of bringing desirable changes in the socio-psychological, economic, and cultural conditions of the rural people. These categories include mainly: (1) the administrators, (2) the supervisors, (3) the specialists, and (4) the field extension workers. In order to achieve maximum effectiveness in performance of the personnel in each of these categories, the importance of training is unquestionable. In fact, people join organizations with the required qualifications, but not necessarily with adequate knowledge and skills for the accomplishment of the work he is supposed to perform. As the situation changes, the people also need to acquire new knowledge, skills, and attitudes to cope up with the changing environment. Therefore, there must be some mechanism by which people in an extension organization can be made more dynamic and responsive to achieve the objectives of helping people help themselves. Training has continued to be considered as the most important device for developing an individual and improving his work efficiency. Through training new knowledge and skills are provided to the individuals in one hand and on the other, existing knowledge and skills are refreshed and sharpened.

An analogy can be drawn regarding training. It is very difficult and laborious to chop a tree with a blunt axe. But if the axe is sharpened, then the job of the chopping the tree becomes easier. Thus, it is always wise to sharpen the axe before using it. Here sharpening of the axe is just like the training of people and working with unskilled personnel is just like the chopping of the tree with the blunt axe.

In fact, training is one of the most important capitals for human resource development for any organization. When the human potential is tapped and put to work, there is the most likelihood that the organization will achieve its goals. However, the nature and mode of training for different categories of personnel in extension organization are likely to different because of their differential nature of job duties and responsibilities. As an approach to employee development, training should work on the assumption that new situation or experience would change an individual at his performance level.

The farmers are the ultimate units who are at the centre of all developmental activities in the developing countries. In the field of agriculture, technology is continuously changing fast in respect of high yielding varieties, fertilizer, pesticides, irrigation, weeding, and management practices. Much skills are needed for use of these technology by the farmers and others concerned in increasing production. Extension personnel and the farmers are, therefore, to be kept abreast with this dynamic agriculture and hence the training of extension personnel and the farmers is of utmost importance (Singh and Srivastava, 1970). For getting desired results in a programme for agricultural and rural development, the field extension workers, the local leaders, and the farmers need to understand the benefits and "know-hows" as well as "do-hows" of improved practices. Hence, it is necessary to arrange timely training programmes to help the extension workers and concerned others acquire necessary knowledge and skills in different aspects of the improved technology. Training to extension personnel is important because it is they who have to teach technology to the farmers.

KINDS OF TRAINING

Training for extension personnel may be broadly classified into two categories: (1) pre-service training and (2) in-service training. Aside from these there are training programmes for farmers, rural youths, women, and local leaders to provide the know-hows and do-hows of tackling problems concerning agriculture and other aspects of their life situations.

Pre-Service Training

The pre-service training refers to the training which is provided to the individuals before entering into the service. Various undergraduate, graduate, and diploma programmes fall under this category. It is offered in schools and colleges through syllabus and curriculum in professional fields. The Block Supervisors (BS) in Bangladesh undergo a 3 year diploma course at the Agricultural Training Institutes before their actual appointment as Block Supervisors in the Department of Agricultural Extension (DAE). Similar training is provided to the Gramsevaks and Gramsevikas in India.

In-Service Training

In-service training are those kinds of training which are provided to the individuals after their appointment in some organization. The various types of in-service training include: (1) induction training, (2) foundation training, (3) on-the-job training, (4) maintenance training, and (5) career or developmental training.

(1) *Induction Training*: Induction training is that having given immediately upon employment to introduce the new staff member to his position (Rogers and Olmsted, 1957). When the employees are freshly recruited by an extension organization, it is quite natural that before starting work they would be eager to know what sort of an outfit they are getting into, what they are supposed to do, and whom they will work with (Van Dersal, 1962). Therefore, the new personnel should be given an orientation about the whole organization he is employed in and other related organizations he will work with. The induction or orientation training is designed for the new employees to acquaint them with the organization, their jobs, their places, and their roles in it. This training is important for future performance of the personnel, because "those things that happen during the first days, the first weeks, or the first few months may determine whether the new extension agent continues to be highly motivated or whether his will to work is weakened or even destroyed."

(2) *Foundation Training*: Similar to induction training, the foundation training is also applicable for the newly recruited personnel. Halim and Ali (1988) states that besides technical competence and routine instruction about organizations, every staff needs some

professional knowledge about some general aspects which may include: leadership ability, communication skill, financial management, administrative capability, acquaintance with various government rules and regulations, etc. This training is proved to the newly recruited personnel during their initial stage of service life so as to strengthen the foundation of their service career.

(3) *On-the-Job Training*: It is a periodic training on how to do the job and adjusting to changes in the nature of work. The immediate supervisors are generally responsible for getting the work done by imparting necessary knowledge and skills and providing adequate guidance. The fortnightly training of the BS under the Training & Visit system of extension is an example of on-the-job training.

(4) *Maintenance or Refresher Training*: This type of training is needed both to keep employees at peak of their possible production and to prevent them from getting into a rut (Van Dersal, 1962). The aim of such a training is to keep the extension workers upto-date with new idea, new information, new methods and techniques, and new developments so as to add to their existing knowledge. It also reviews the older materials, ideas, and practices that are still workable. Various kinds of short courses, seminars, workshops, and conferences may fall under this category of training.

(5) *Career or Developmental Training:* It is designed to increase the knowledge, skills, and aspirations of the employees to make them ready to accept greater responsibility when opportunities are available.

Farmers' Training

In the process of technology transfer, the training of farmers, farm leaders, rural youths, and women plays an important part. Farmers' training is a process of non-formal education which educate the farmers in a group on some specific topic at a particular place. A group of farmers, from leaders, women, or farmers' sons are brought to a training centre and are provided training for a certain period on some specific subject-matter. The farmers can also be trained on some technology right at their home or farm through regular visits by the extension workers. The Block Supervisors under the T & V system trains the contact farmers by delivering carefully prepared impact points though their fortnightly visits.

APPROACHES AND METHODS OF TRAINING

Changes of Training Approaches

During the recent years, there has been a remarkable departure from the traditional way of conducting training sessions. The departing point began from the realization of the very approaches to child learning and adult learning processes. Knowles (1970) has classified the differences that emerge in the learning process as individuals gain maturity. He pointed out that the traditional teaching method heavily relies on the technology of "pedagogy"—which means, specifically, the art and science of teaching children. In contrast, "Andragogy" a term which combines the Greek stem *andr* (meaning 'adult') with agogos" (meaning 'leading') is defined as the art and science of teaching adults. Following are the four basic concepts which shows difference between Andragogy and Pedagogy (Ali and Islam, 1989):

a. *Self-concept:* The self concept of a child is that of being a dependent person. As children move toward adulthood, they become increasingly aware of being capable of making decision for themselves. This change is generally known as psychological maturity and the adult gradually tend to be autonomous in decision-making. Because so many of our training environments have been influenced by traditional pedagogical practices, adults come to attend training programmes expecting to be treated like children. When adults discover that they are capable of self-direction in learning too, they often exhibit a remarkable increase in motivation to learn.

b. *Experience:* Adults in course of living, have gathered large amount of experience of various kinds. Children have relatively little experience on which to reflect. In the andragogical approach to learning, the experience of adults is valued as a rich resource for learning. In the tradition of pedagogy, the tendency has been to regard the experience of children as being of little worth in the learning process. It is possible because of this reason that the methodology of pedagogy has been largely oriented toward "one way communication techniques", i.e., lectures, assigned readings, and audio-visual presentations. Andragogy, on the otherhand, is full of "experiential"—a term used by McCaffey (1984). It includes two way and multi-directional techniques like—group discussion, simulation and role-playing, team building, case study, skill practice, and so on.

c. *Readiness to Learn:* The prime factor responsible for effective dissemination of skills and knowledge is the readiness of the learners to learn. This could largely be ensured by involving the learners in selecting the learning contents. In traditional pedagogy, the teacher decides the contents and himself assumes the responsibility for how and when the learning will occur. In andragogy, learners are grouped according to their individual interests and learning needs. The learners decide what they want to learn based on their own perception of the demands of their social/field situation. The 'facilitator' of adult learning acts as a resource person just to help the learners form interest groups and diagnose their learning needs.

d. *Time Perspective and Orientation to Learning*: Adult learning is "problem centered" rather than "subject centered". Unlike child learning, it is not "preparation for the future" rather is a process for problem finding and problem solving in the present. To discover "where we are" and "where we want to go" is the primary concern of the andragogical approach to learning. One point, however, may be made clear with regard to andragogic and pedagogic approaches. They are not to be considered mutually exclusive. Some overlapping may be necessary to achieve maximum results depending on the situation.

METHODS OF TRAINING

Various instructional methods can be used for conducting training programmes for extension personnel and farmers. No single method will satisfy the needs of a training programme. However, the using of correct methods will depend to a large extent on the objectives of the training and the specific conditions in which learning is to take place. During the last twenty years, many instructional methods have been developed, tested, combined, and adopted to different learning situations. As discussed earlier, the emphasis is shifting from the pedagogical nature to the andragogical nature for effective training. There are some methods which are simple and can virtually be used by any teacher/trainer without any preparation, while others are fairly sophisticated and need extensive preparation of both the trainers and the course participants. However, the following are some of the commonly used instructional methods.

The Lecture/Classroom Presentation

Formal classroom training is probably the most familiar and oldest method of instruction. Lectures, movies, and exercises are the most commonly used methods of conducting classroom training. The lecture method is suitable where: (i) large number of participants are to be trained, (ii) scarce teaching resources are available, and (iii) a few hours are available for instruction. This method is, however, unsuitable for teaching skills since it is mostly characterized as a one-way communication.

The Discussion

Structured discussions are conversations between trainees, aimed toward specific learning objectives (Laird, 1985). Through discussion, ideas and opinions on a particular subject are freely exchanged among the trainees themselves as well as between the trainees and the instructors. This helps in obtaining feedback from the participants. There are various forms of conducting discussions, some of which are: panel discussion (also called colloquies or symposia), structured discussions, panel discussion by students, topical discussion, question-answer panels, cognates, and open-forum discussions (Laird, 1985). In these discussions, the learners usually listen, watch, read, move, write or respond in relation to the specific stimulus presented. The discussion methods help participants change attitudes after discussions. However, the participants may stray from the subject-matter or fail to discuss it usefully and thus the whole session may be blurred and boring.

Reading Assignments

In this method the participants are given reading assignments of books, hand-outs, or any other reference materials. To make the reading assignments more meaningful, some kind of accountability must be established. The learners not only goes over the pages of the material but also comprehend the contents for application to real-world situations. The reading assignments may be followed-up by discussion, participation on a panel of students, or application to a simulated problem.

Demonstration

These are illustrated lectures or presentations. The lecture is supported by specimen, models, and various types of visual and audio-visual aids. Here the participants get an opportunity to see, hear, and do a practice being presented in the training. The key to successful demonstration is a close integration of the spoken and visual stimulus. Hopefully, they should be simultaneous. To make demonstration an effective teaching method, careful planning and rehearsing is necessary.

Field Trips

If properly managed, field trips can be a very good method of learning. The instructor sets up expectations and objectives before the trip takes place. The participants are given certain questions on some topic(s) for which they have to find answers while they are taken to the field trips. After the field trip, a feedback session is held to review the answers and observations of the participants. This method permits learners to experience sensory impressions which could never occur in classroom or conference rooms.

Brainstorming

This is a specialized form of discussion where each participant is stimulated to express his ideas and suggestions on a particular problem or issue. The learners suspend their judgements until a maximum number of ideas have been generated. When ideas from all participants are obtained, these are critically evaluated and the best ideas are accepted. For selecting the best ideas, each idea is tested against certain pre-established criteria. This method is based on the synergistic theory that groups can generate more ideas and better ideas than all the individual members could do if they worked independently (Laird, 1985).

Case Studies

The participants are supplied with a printed description of particular problem situation. They are to read the case, analyze it, and suggest appropriate recommendations so as to overcome the problem. The case study can be an effective teaching-learning mechanism if the cases are developed typical of the extension organisation for which learning is designed. In andragogic approach, the instructors encourage the participants to write their own cases from real life problems. Threadbare discussions are held, preferably in smaller buzzgroups, about each case provided or generated.

Role plays

In roleplays the participants are asked to re-enact the situation which they face on the job, or which they will face in the future, or which they perceive to be job-like (Laird, 1985). This method is suitable for the practice of dealing with face to face situations. It adds involvement, variety, reality, and specificity to the learning experience. To make the role plays totally relevant and realistic the instructors sometimes ask participants to write up their own roleplays. However, one disadvantage of the roleplays method is that the participants may be embarrassed and their confidence may be sapped rather than build up.

Simulations and Educational Games

Here the participants are given some situation and asked to find out solution through the exploitation of their brain initiative and drive (Halim and Ali, 1988). They discuss the critical situations and make their decisions. This method is, thus, suitable for the training of decision-making. It encourages active participation of the trainees and stimulates intelligence and power of thinking. Small educational games add motivation to learning and give insight to "going outside the problem in order to solve the problem". This method, however, is not appropriate for lower level participants. The participants may not take the training situation seriously.

Modular or Self-learning Model

In this training method, the participants are provided with the training materials in the form of modules. Each trainee learns on his own from these modules. A feedback system is generally incorporated in the module such that the trainee can evaluate his own performance. This method is easy to administer and require minimum supervision. However, the participants may not be motivated to use this method if it is not associated with any reward system (UNESCO, 1988).

Other Methods

Some other important teaching techniques used in training of personnel include: buzzing, action mazes, card sorts, problem clinic, advisory committee, task forces, conferences, seminars and workshops, in-basket exercises, etc. Most of these techniques need special skill and experience for their proper planning and exercise.

ESSENTIALS OF A TRAINING PROGRAMME

Major Components of Training

In organizing a training programme for the extension personnel or the farmers, at least five aspects must be considered. These are: the trainees, the trainers, the subject-matter or course content, the teaching methods, and the resource allocation (fund, space, equipment, furniture, teaching aids, vehicles, étc.).

The Trainees: Proper selection of trainees is important. Since a training is objective - oriented, its success depends to a greater extent on how the participants respond enthusiastically to the programme. In the selection of trainees, therefore, homogenity or heterogenity of participants should be considered. If the participants have heterogeneous interest and eagerness, the interaction would be less and the training output is likely to be minimum. On the other hand, more fruitful interaction will result in a homogeneous group of participants as they have similar needs, aims, and objectives and they come from almost similar organizational strata. Therefore, while selecting trainees similarity in terms of qualification, job responsibility, and other related factors should be kept in mind. The trainer: The trainer must be competent enough to handle the course content efficiently. He should have skills in using most of the teaching techniques and should possess sufficient field experiences. For successful accomplishment of the training programme, the trainers should be informed quite well ahead about the backgrounds (qualification, nature of job, etc.) of the participants such that he can relate his discussion with the day to day business of the trainees. It is necessary that the trainer creates a free, friendly, fearless and democratic atmosphere in the class such that effective communication takes place between the trainer and the trainees. It is also advisable that the trainer first prepares properly the minds of the trainees to receive the instruction and he should possess the quality of empathy.

The Content or Subject-matter: For every training there must be a content or subject-matter on which the trainer will impart training to the trainees. The participants will learn, understand, and remember certain information through the training process. If a training is to be effective in performing the job by the trainees, then the content of the training must be based on the needs of the trainees. That is why training need assessment activities must invariably be taken up before organising any training programme.

The process of training need assessment involves atleast three major considerations: the job analysis, the individual analysis, and the organizational analysis. Objectives of a training programme should commensurate with the needs of the organization and/or the employees. Therefore, before setting the objectives, need analysis is very much important.

The Teaching Method: The training programme cannot be effective if appropriate teaching methods are not used. It has been mentioned before that there are various types of teaching methods, which may vary from low to high in terms of the level of learner participation. Learning will be more effective if greater participation by the trainees is ensured. Therefore, selection of appropriate methods and techniques is very much important. However, what method to use will depend on a number of factors such as: learning objectives, backgrounds of the learners, knowledge and skills of the trainers, the facilities available, etc.

The Resource Allocation: The other elements of a training programme cannot function properly if there is no adequate resource allocation in terms of finance, physical facilities, equipments, visual and audio-visual aids, transport for mobility, and other logistic supports. The creation and management of a learning situation conducive to effective learning for the participants will be difficult if there is no easy arrangement for supply of the required resources.

Essential Elements of a Training Schedule

A training schedule is a plan which shows when, how, and by whom different activities related to a particular training programme will be carried out. Each formal training programme usually has a training schedule which has the following elements:

1. The title of the training course
2. The objectives of the course
3. The participants or the trainees
4. The duration of the course
5. The coordinator of the course
6. The sponsoring agency of the course
7. A calendar of operation showing the date, time, topic/event, place, and the teachers/instructors.

Aspects of Evaluation

Various aspects are involved in a training programme. The evaluation of a training programme should include the following major aspects:

1. The effects of the training programme
2. The methods, techniques, and media used in the programme in order to achieve the objectives
3. The trainers/instructors
4. The trainees
5. The content of training (what has been taught)
6. Overall management of the training programme.

1. *The Effects of Training:* The impact of a training should be reflected in the change of job behaviour of the participants before and after training. But the effects are so varied and multi-dimensional that the assessment of the changes of job behaviour of the participants after training becomes the most difficult task. This is particularly so on the ground that behaviour of an employee in a job situation might be affected by a number of factors other than the training programme attended. A few of them could be higher pay, advancement in science and technology, recognition of work by the superiors, better working environment, and so on. Since, it is neither possible to include nor control all the factors in one evaluation, the observations and judgements of the immediate groups who are directly concerned with the participants, can be a better approach for assessing the behavioural and attitudinal changes of the participants after training. The immediate groups may consist of the superiors, the peers, and the sub-ordinates and/or farmers. The opinions of these immediate groups can be assessed through a guided discussion with them on preselected areas of job performance of the trainees. In this approach, however, there might be some weaknesses such as the immediate groups all may not observe the participants both before and after (because of transferable nature of their job), the subordinates may not feel free to express the negative observations about their superiors, and the interest of the grader may sharply decrease if the number of persons to be graded at a time increases greatly. As a whole, the impact of a training programme should be judged on the criteria of (i) knowledge gained, (ii) knowledge applied in the field, and (iii) the results obtained in terms of increase in productivity.

2. *TheMethods and Techniques Used in Training:* The methods of presentation of the subject-matter may vary according to situations. It is necessary to evaluate what method(s) is/are effective in transferring knowledge, skills, and attitudes such that the objectives of the training are achieved and the job performance level of the participants is increased to the maximum. The basic question to be answered in the evaluation of methods is: "To what extent did the methods used in training assist in transferring the subject-matter, so that it was understood, learned, remembered, and used," (Chitamber, 1961). Although an experimental design (controlled research) is a better way to determine the effectiveness of a particular teaching method or technique, a qualitative judgement can be assessed by asking the participants in a post-training situation to indicate how effective they felt of the different methods and techniques used along a scale, e.g., "very effective," "effective", "fairly effective", "not effective", and "not at all effective".

3. *The Trainers/instructors:* The success of a training programme also depends to a large extent upon how effectively the teachers and instructors present their subject-matter to the participants. A teacher's effectiveness can also be evaluated in terms of several dimensions such as his: (i) knowledge, interest and the technical soundness of the subject-matter taught, (ii) method of presentation, (iii) applicability and relevance of subject-matter to practical field problems, (iv) relationship with trainees, colleagues, and others, (v) sense of responsibility and dependability, and (vi) general manner, bearing and appearance (Chitamber, 1961). Against each dimension a scale showing "very good", "good", "fair" "poor" and "unable to observe" is usually used. The participants are the central people who give their ratings about the instructors according to the above scale. Aside from the participants, the Director or Principal of the training institute/centre, the colleagues, and the relevant experts can also rate the instructors by using the same rating scale.

4. *The Trainees:* Evaluation should also be done about the trainees or participants themselves. The aspects of trainee evaluation can be their: (i) knowledge and understanding about the programme and the subject-matter, (ii) skills in using extension techniques and methods, (iii) interpersonal relationships with colleagues and others, (iv) adjustment to new situation, (v) interest and capacity to develop themselves, (vi) punctuality, and regularity, and (vii) extent of interaction in the class. The instructors will usually rate the trainees

using the rating scale (very good, good, fair, poor, and unable to observe) for each aspect. The overall performance of the trainees can be ascertained by employing appropriate tests.

5. *The Content of Training:* The content of training or what has been taught is of prime importance for any training programme. Some of the questions to be raised in this connection are: (i) to what extent the subject-matter taught in the training are relevant and useful in performing the job of the participants? (ii) which of the topics are most useful and which topics, least useful? (iii) what topics should be deleted from the training programme and what additional topics should be included to make the training more useful and interesting? Appropriateness of the content should be judged in terms of the applicability of the knowledge and skills gained through the training.

6. *Overall Training Management:* Whether the learning experience of the participants will be satisfying or dissatisfying will depend to a great extent on what impressions they get regarding certain management aspects such as providing facilities for classroom, use of visual and audio-visual aids, supply of teaching materials. library, food and accommodation, recreation, transport, etc. The impression of the participants can be evaluated by asking them to indicate how adequate was each of these aspects of the overall training management. The scale for measuring adequacy could be "very adequate", "adequate", "fairly adequate", "not so adequate", and "not at all adequate". The participants' feeling about the whole management of the training can also be assessed by asking them to point out things of their most liking and the things of most disliking that they felt from the very beginning to the completion of the training programme. This is, in fact, another way to know the grievances of the participants in order to improve the future training programmes. The participants should be given opportunities to express their open suggestions for improvement of the quality of training in future.

CONCLUSION

Training of both the extension personnel and the farmers is an useful tool for technology transfer in agriculture and/or rural development. It is a costly investment and must be planned, budgetted, and evaluated with care. In agricultural extension, farmers are the key elements that must be changed and in doing so the role of technology transfer can never be overemphasized. Personnel at various levels in an agricultural extension organization need to be equipped with adequate and up-to-date "know-hows" and "do-hows" of dealing with the problems of the farm people. Appropriate training programmes for these personnel will definitely contribute significantly towards their development in knowledge, skills, and attitudes. But mere providing training will not help much in bringing desired change to the people unless proper arrangements and environments are created for the application of the knowledge and skills gained by the participants through the training programmes.

The transfer of technology being a complex and continuous process, there is no alternative but to develop the human resources to achieve desired goals. The effective manager of extension organizations are those who can best develop all human resources (including farmers) involved in planning, implementing, and evaluating extension programmes for the purpose of achieving organizational objectives. In this respect, training must be considered as a continuing function and the process of training and retraining of the available human resources must be an integral part of any organizational programme.

References

Ali, M. Hazrat and M. Monirul Islam. 1989. *Evaluation Report the Training on Management of Development Programmes for the Upazila Level Officers under NRDP-II/DANIDA*. Graduate Training Institute, BAU, Mymensingh, Bangladesh.

Chitamber, J.B. 1961. "Evaluation of Training Programmes", in Kamath, M.C. (ed). *Extension Education in Community Development*. New Delhi: Directorate of Extension, Ministry of Food and Agriculture, Govt. of India.

Flippo, E.B. 1961. *Principles of Personnel Management*. New York: McGraw-Hill Book Company, Inc.

Halim A. and M. Mozahar Ali 1988. Administration and Management of Training Programmes. *Bangladesh Journal of Training and Development*. 1(2): 1-19.

Kirkpatrick, Donald L. 1983. *A practical Guide for Supervisory Training and Development*. 2nd ed. Reading, MA: Addison-Wesley Publishing Company, Inc.

Knowles, Malcolm S. 1970. *The Modern Practice of Adult Education*. New York: Association Press.

Laird, Dugan 1985. *Approaches to Training and Development*. Reading: Addison-Wesley Publishing Company, Inc.

McCaffey, J.A. 1984. *Independent Effectiveness: A Reconsideration of Cross-cultural Orientation and Training*. Alexandria, VA: Training Resource Group.

Nadler, Leonard. 1970. *Developing Human Resources*. Houston: (Supported by the National Society for Training and Development).

Rogers, F.E. and A.G. Olmsted. 1957. *Supervision in the Cooperative Extension Service*. Madison, Wisconsin: National Agricultural Extension Centre Advanced Study.

Singh, Daulat and V.K. Srivastava. 1970. "Training of Extension Personnel and Farmers in Uttar Pradesh", in Singh, K.N., C.S.S. Rao, and B.N. Sahay (ed). *Research in Extension Education for Accelerating Development Process*. New Delhi: Indian Society of Extension Education.

Tylor, M.H. 1961. "Training of Trainers". Kurukshetra. Cited in Dwarkinath, R. "Training of Extension Personnel and Progressive Farmers", in Singh, K.N., C.S.S. Rao, and B.N. Sahay (ed). *Research in Extension Education for Accelerating Development Process*. New Delhi: Indian Society of Extension Education.

UNESCO. 1988. *Manual on Cost Effectiveness of Training Modalities in Population Education*. Bangkok: UNESCO Principal Office for Asia and the Pacific.

Van Dersal, W.R. 1962. *The Successful Supervisor*. New York: Harper and Row Publisher.

18

Training of Farmers for Effective Technology Transfer

Ranjit Singh

There is no doubt that agricultural production has increased manyfold during the last 25 years, yet a lot more needs to be done. The population in our country is increasing at an alarming rate and it will cross the 1000 million mark in the year 2000 AD. The foodgrain needs of this size of population have been estimated by the National Commission on Agriculture of the order of 220 million tonnes. We are indeed far away from the projected requirement. Continuous expansion in agricultural production in general and foodgrains, pulses and oilseeds in particular, has to be maintained as our goal for the future. This is to be realised through an emphasis on further improvement in the productivity of major crops. The success in increasing agricultural production depends upon the ability to involve a large number of farmers, big and small, in the production efforts. It has been observed that the benefits of new technology and of economic development took longest to reach the poor communities and contributed least to improve the quality of their lives.

For motivating farmers specially small farmers to adopt new technology, practical training is a must. In fact, practical training is an essential condition of any rural development scheme. Training of farmers through organised courses can be considered to be an intensive training activity which is carefully planned and properly implemented with a view to educate the participants systematically and thoroughly about a pre-determined subject-matter. As such a farmer's course in contrast to either individual contact or mass method of teaching, is a formal group method of teaching so as to impart knowledge, teach new skills, and bring about desirable changes in the behaviour of the participants.

According to Lynton and Pareek (1973) training, as we see it, aims at a lasting improvement on the job. The kind of education we call training—more of this distinction later—is in truth not for knowing more but for behaving differently. Lewin (1947) describes development as a succession of three stages 'unfreezing' 'moring' and 'refreezing'. A successful training programme is that which can lead the training from one stage to the other. Miles (1964) conceptualizes this process as a change—inducing temporary system. The guiding principles for working out training strategies lie in these dynamics of development process.

When to Organize Farmers' Training

Organising the training will depend upon certain peculiar pre-requisites that determine the possibilities whether or not training courses can be held effectively. Presence of these pre-requisites are essential for the success of training programme.

1. Availability of sound technology that is superior to the prevalent practices.
2. A realisation among the farming community that superior technology exists and they can profit by acquiring it.
3. Qualified and trained personnel who can successfully transmit new technology to the participating farmers so that they can grasp it and employ it at their respective farms.
4. A desire on the part of the farmers to get personally acquainted with the specialists who can be depended upon in the time to come for advice regarding the specialised topics.
5. There are increasing number of problems that can be solved by the specialists rather than a generalist.
6. There is a large segment of clientele which is neither affluent enough to take initiative to obtain information individually at their own initiative, nor persons who will never care to move out of their places to pick any ideas. It is neither the innovators nor the laggards who will represent the potential participants of any training course.
7. In view of the above pre-requisites, it can be hypothecated that a certain amount of maturity of extension services in particular area coupled with an exacting farming community, that it seeks to serve, is necessary for the farmers.

Objectives of Training

1. To attract the genuinely interested persons who are either self-motivated or externally stimulated to take part in the course.
2. To impart new knowledge, teach better skills and bring about the desirable changes in their behaviours. There must preferably be no incentive to the participants other than the ideas and information that they can obtain from these courses.
3. To place the latest knowledge in the hands of desirous persons.
4. To develop acquaintance between the farmers and specialists and also the scientists.
5. To encourage mutual exchange of experience among the participating farmers.
6. To pave way for the flow of problems of the farmers to the research centres.
7. To enhance faith in the research findings.

Courses can be of two types (a) general courses, and (b) specialised courses. General courses dealing with crop husbandry can be held twice every year before each crop season for about one week's topic, like horticulture, plant protection, dairy, poultry, etc. Long-duration courses, especially for the young farmers, can also be organised where basic theory of agricultural science with emphasis on practical operations is imparted to educated young persons who have decided to adopt farming as their profession.

Essentials of an Effective Course

To make a training course successful, care should be taken of the following:

1. Adequate and timely publicity of the date, duration and contents of the courses so as to attract those having the potentiality and desire to profit by a course.
2. An optimum number of participants. A good size seems to be anything from 25 to 40 wherein each and every individual can be expected to benefit personally.
3. Proper orientation of the participants on the opening day so as to pin down their attention upon the need of extraction the best results in the form of new ideas and information applicable to their farm situation.

4. Training must be accompanied by immediate opportunities for farmers to try the new things they have learnt.
5. If possible, training should be organized where the farmers are, i.e. nearer their farms and homes.
6. A balanced programme which is of interest and relevant to participants.
7. Provision for discussions, demonstrations and field trips to be planned one after the other. Use of variety of teaching methods promote learning.
8. Continuity of the programme so that participants do not remain un-occupied at any time.
9. Provision of supplies on a limited scale.
10. Effective communication by the speakers along with active involvement of trainees.
11. A trip to places of educational and historical interest.

Characteristics of Participants

It will be beneficial for an organizer of the course to be conversant with the qualities that he may seek among the participants. A few of these are as under:

1. The participants of the course need not be innovators or laggards. It will be too much to expect from the innovators to stay in a course for an extended period. The laggards may not be in a position to conceptualise and comprehend the newer innovations that they may not have seen earlier.
2. A great majority of them must have a desire to learn.
3. They should be homogeneous as a group so far as their level of interest, background experience, education and expectations are concerned.
4. Preference should be given to literate persons.
5. They should be practising farmers.
6. They should value the new ideas and information that they are able to get as the result of their participation in the course.
7. They should be willing to share their experiences with others.

Improving Farming Training

a. Training must reach the farmers where they are. Extension agencies, agricultural colleges and universities invite farmers for training in their institutions. The attendance of the farmers in training programmes specially organised in training institutes is always very thin and the participating farmers also feel that many innovations demonstrated on institutional farms may not be successful at their own farms. To create better conviction in the mind of the farmers for innovations and also not to disturb their own programmes, it is always feasible that the training must be organised in villages.

b. Training must be directed, especially to farmers' present interests and needs. How to increase crop and livestock production and how to increase profit? Money is the best incentive. With the introduction of high yielding varieties the farmers have very well understood that training relating to agriculture production if it does not result an increase in his income, it is waste of time for him. Therefore, the training programme must be directly related to the farmer's monetary interests.

c. Training must accept the fact that the farmers are adults. They are not children and should not be trained in the way children are trained.

d. Training must be fitted into period when the farmers are not too busy, during night or off-season.

e. The subject for training must be of a particular new or changed practice—what is it? Why it is superior? How can be carried out?

f. Training must be accomplished by immediate opportunities for farmers to try the new things they have learnt—application of fertilizers, use of supplies and equipment and also make things available to them.

g. Each new or changed practice must be technically sound and economically profitable.

h. Farmers need encouragement to try to do things—acquiring knowledge and skill is not enough, they also need encouragement to try out new practices.

i. Training should be imparted in those technical subjects which can be used by the trainees in the near future.

j. There should be proper arrangement for conducting follow-up.

Effective Teaching Learning Situation

In a training course, the trainer has to create an effective learning situation. A good learning situation is one in which the individual desires to act or to solve a problem which blocks his progress. It is generally known that learning takes place best in an environment which is attractive, pleasant and comfortable. Learning situation should encourage and stimulate the trainer. The quality of learning is evident from the type of behavioural changes accomplished by the learner. The learner is the central figure in the teaching and learning. In case he is unwilling and gives a negative response to various stimuli, the desirable change in his ideas and behaviour can hardly be achieved. In extension education the learning process includes both obtaining and retention of knowledge. If the physical settings like place, light, ventilation, seating arrangements, etc. are not provided properly, effective teaching and learning cannot take place. There should not be any outside distraction. In addition to have comfortable seating the arrangement of seating should be in such a way that every person can see and hear every other person easily. Adequate reading material also needs to be provided. Learning will only take place if learner really wants to learn and ready to change. More learning will take place when the learner realizes the importance of learning. Training should start where the learner is. Use words that are simple and familiar and at a speed which provides sufficient time to the learner to grasp and absorb, what has been said. An informal environment is more condusive for adults to learn. The teacher should guide and stimulate rather than dominate the group through observing rigid classroom rules. Involvement of emotions accelerates the speed of learning. Trainees should obtain constructive emotional involvements without providing excessive and destructive irrationality.

Training is the acquisition of the art of utilization of knowledge and skills. As pointed out by Solon Barraclough (1973), effective training is not a process of cramming information into the heads of trainees, but consists of a real interaction between students and teachers. In order to achieve it, we should avoid four imbalances while formulating the programme (Miles, 1964). These are input overload, unrealistic goals, alienation and linkage failure. True learning will take place only if a learner feels some change after his exposure to a learning experience. This change may be in knowledge, attitude or skill. Best results will be obtained when there is desirable changes in knowledge, attitude and skill after the completion of training.

References

Coombs, P.H., Ahmed, M. 1974. Attacking Rural Poverty, Baltimore, The John Hopkins Uni. Press.

Fisher, J.D. *et al.*, 1963. Agricultural Extension Training Nairobi, USAID.

Lewin, K. 1947. Group Decision and Social Change. New Comb & E. Hardly Teds, Reading in Social Psychology, NV.

Lynton, RP & Pareek U. 1973. Training for Development, Bombay, D.B. Taraporevala Sons & Co.

Maunder, A.H. 1972. Agricultural Extension—A reference Manual, Rome FAO.

Miles, M.B. 1964. Innovation in Education. N.Y. Teachers College, Columbia Uni, pp. 437-90.

Mosher A.T. 1966. Getting Agriculture Moving. N.Y. Harpar & Row.